Lecture Notes in Mathematics

Edited by A. Dold and B. Eckr

734

Ring Theory Waterloo 1978

Proceedings, University of Waterloo, Canada, 12–16 June, 1978

Edited by David Handelman and John Lawrence

Springer-Verlag
Berlin Heidelberg New York 1979

Editors

David Handelman
University of Ottawa
Ottawa, Ontario K1N 6N5
Canada

John Lawrence
University of Waterloo
Waterloo, Ontario N2L 3G1
Canada

AMS Subject Classifications (1970): 06A60, 13D99, 16A26, 16A40, 16A46, 16A48, 16A50, 16A54, 16A62, 18F25, 46L05

ISBN 3-540-09529-2 Springer-Verlag Berlin Heidelberg New York
ISBN 0-387-09529-2 Springer-Verlag New York Heidelberg Berlin

Printed in Germany

Printing and binding: Beltz Offsetdruck, Hemsbach/Bergstr.
2141/3140-543210

RING THEORY, WATERLOO 1978

A ring theory conference took place at the University of Waterloo, 12-16 June 1978, and these are its proceedings. This conference was held as a part of the Summer Research Institute in Ring Theory, at Waterloo, sponsored by the Canadian Mathematical Society. A list of talks given by Members of the Institute is given on page 1 .

In soliciting speakers, and contributors to the Proceedings, we attempted to represent those portions of ring theory which seemed to us interesting. There was thus considerable emphasis on lower K-theory and related topics, artinian and noetherian rings, as well as actions and representations of groups on rings. Regrettably, we could only obtain one paper in the mainstream of commutative ring theory, but we believe that the lack of quantity is more than made up for by the quality.

We have an article by A.A. Suslin (communicated by R.K. Dennis) which elaborates on his talk at the Helsinki International Congress of Mathematicians, as well as a paper by K.R. Goodearl, on recent results of Zaleskii and Neroslavskii previously available only in Russian.

We also took the liberty of including a survey of results in a field which we feel deserves more attention by ring theorists, C*algebras from an algebraic point of view.

We would like to thank, in addition to the speakers, contributors and extremely efficient referees (all papers were subject to refereeing),

Mme. Lucie LeBlanc (University of Ottawa), and Mrs. Sue Embro, Ms Rose Scherer (University of Waterloo). Finally we would like to thank the National Research Council of Canada, the University of Waterloo, and Professor W. Forbes, Dean of Mathematics at the University of Waterloo.

David Handelman
University of Ottawa
Ottawa, Ontario K1N 9B4
Canada.

John Lawrence
University of Waterloo
Waterloo, Ontario N2L 3G1
Canada.

TABLE OF CONTENTS

Summer Research Institute Workshop Members

Name	University
W. Burgess	University of Ottawa
R. Burns	York University
A. Carson	University of Saskatchewan
D. Handelman	University of Ottawa
A. Heinicke	University of Western Ontario
K. Nicholson	University of Calgary
K. O'Meara	University of Canterbury (New Zealand)
G. Renault	Université de Poitiers

Talks Given in the Workshop of the Summer Research Institute

(all talks were about 1 hour in length).

Name	University	Title of Talk
J. Lawrence	University of Waterloo	Relations in regular rings
K. O'Meara	University of Canterbury	Maximal quotient rings of prime group algebras
D. Handelman	University of Ottawa	The natural orderings on $K_0(R)$
K. Nicholson	University of Calgary	Normal radicals and Morita Contexts
A. Carson	University of Saskatchewan	$\aleph_0$ categorical biregular rings.
V. Dlab	Carleton University	Recent results in abelian groups
J. Lawrence	University of Waterloo	Continuous rings and $\aleph_0$ continuous rings
A. Heinicke	University of Western Ontario	Non-commutative analogues of regular local rings.
D. Handelman	University of Ottawa	$\aleph_0$-continuous rings and affine functions on Choquet simplices.
G. Renault	Université de Poitiers	A survey of group actions on rings.

RING THEORY CONFERENCE

11-15, June 1978

LIST OF PARTICIPANTS

A. Bak, Universitat Bielefeld
H. Bass, Columbia University
W. Burgess, University of Ottawa
R. Burns, York University
M. Chacron, Carleton University
A. Carson, University of Saskatchewan
J. Dauns, Tulane University
B. Dayton, Northeastern Illinois University
K. Dennis, Cornell University
V. Dläb, Carleton University
G. Elliott, KØbenhavns Universitets and University of Ottawa
P. Fillmore, Dalhousie University
J. Fisher, University of Cincinnati
P. Fleury, State University of New York, Plattsburgh.
K. Fuller, University of Iowa
M. Gabel, Purdue University
S. Geller, Purdue University
K. Goodearl, University of Utah
J. Goursaud, Université de Poitiers
D. Grayson, Columbia University
P.L. Gupta, University of Maine
D. Handelman, University of Ottawa
A. Heinicke, University of Western Ontario
M. Hochster, University of Michigan
R. Irving, Brandeis University
S.K. Jain, Ohio University
R. Kruse, St. Mary's University
J. Lawrence, University of Waterloo
L. Levy, University of Wisconsin
K. Louden, Concordia University
G. Mason, University of New Brunswick
K. McDowell, Wilfred Laurier University

List of participants cont'd....

K. Murasugi, University of Toronto
K. O'Meara, University of Canterbury
J.D. O'Neill, University of Detroit
B. Osofsky, Rutgers University
J. Osterburg, University of Cincinnati
J. Park, University of Cincinnati
R. Raphael, Concordia University
J. Rätz, University of Bern, Switzerland
G. Renault, Université de Poitiers
L. Roberts, Queen's University
P. Roberts, University of Utah
D. Saltman, University of Chicago and Yale University
C. Sherman, New Mexico State University
R. Snider, Virginia State University
J. Sonn, Technion, Haifa, Israel, and McMaster University
T. Stafford, Brandeis University
R. Varley, University of Utah
R. Warfield, University of Washington
C.A. Weibel, IAS , Princeton

Ring Theory Conference Talks

12 June:	K.R. Goodearl	University of Utah	The state space of K_0 of a ring
	J. Dauns	Tulane University	Noncyclic crossed product division algebras
	T. Stafford	Brandeis and Cambridge	K-theory and stable structure of noetherean group rings.
	S.K. Jain	University of Ohio	PCQI rings and a question of Faith.
13 June:	R.K. Dennis	Cornell University	A survey of K_2
	M. Hochster	University of Michigan	Principal Ideal theorems
	G. Renault	Université de Poitiers	Actions de groupes et anneaux réguliers injectifs
	B. Osofsky	Rutgers University	Projective dimension
	J. Goursaud	Université de Poitiers	Sur les anneaux réguliers.
14 June:	K. Fuller	University of Iowa	Biserial rings
	H. Bass	Columbia University	Representation of infinite groups
15 June:	R. Snider	Virginia State University	Is the Brauer group generated by cyclic algebras?
	H. Levy	University of Wisconsin	Modules over the cyclic group of prime order.
	D. Saltman	Chicago and Yale	Reduced norm of division algebra.
	G. Elliott	Copenhagen and Ottawa	Ordered groups
	S. Geller	Purdue University	Excision for SK_1
16 June:	J. Fisher	Cincinnati University	Semiprime crossed products.
	L. Roberts	Queens' University	Counter-examples to excision for SK_1
	R. Warfield	University of Washington	The stable number of generators of a module.

On totally ordered groups, and K_0 [†]

George A. Elliott

Mathematics Institute, University of Copenhagen

Abstract. Some results are described concerning totally ordered abelian groups. These can be interpreted, via the functor K_0, as classification results for certain noncommutative rings, for which K_0 as an ordered group happens to be a complete invariant.

This work was done while the author was a Guest Scholar at the Research Institute for Mathematical Sciences, Kyoto University, partially supported by a grant from the Carlsberg Foundation.

[†] Formerly titled, "On totally ordered groups" .

1. Introduction. This paper derives a new property of totally ordered abelian groups (see section 2).

In [10], a countable ordered abelian group was called a dimension group if it could be expressed as the inductive limit of a sequence of finite ordered group direct sums of copies of $\mathbb{Z}$. Theorem 2.2 shows that any countable totally ordered abelian group is a dimension group.

Dimension groups were introduced in [10] for the purpose of classifying certain noncommutative rings. It happens that the abelian group K_0 of the inductive limit of a sequence of semisimple matrix algebras (over a fixed field), when ordered in a natural way, is a dimension group. Moreover, every dimension group arises in this way, from such an inductive limit ring, and this ring is in fact unique up to Morita equivalence.

It is therefore interesting from the point of view of ring theory to study certain aspects of the theory of ordered groups, and of totally ordered groups in particular. Some modest contributions to this theory are described in sections 3 and 4 of this paper. Some comments on the ring-theoretical interpretation of these results are made in section 5.

The main result of sections 3 and 4 is that a dimension group with the decreasing chain condition on ideals, and with simple subquotients isomorphic to $\mathbb{Z}$, is determined by its lattice of ideals (or by its spectrum). The first step is to show that the order is determined by the ideals (together with the orientation of the simple subquotients) (3.6). In particular it is shown that if the lattice of ideals is totally ordered, then so is the group (3.9). The second step, easy enough in the totally ordered case, is to obtain a decomposition of the underlying group into a direct sum of copies of $\mathbb{Z}$, compatible with the ideal structure. It follows that this is in fact a decomposition of the ordered group into a lexicographical direct sum of copies of $\mathbb{Z}$, and computation shows that the index set in such a decomposition is in a natural way order isomorphic to the set of prime ideals, i.e., the spectrum.

A few words on terminology are in order. Only abelian groups will be considered. Such a group G is said to be ordered if there is given a semigroup $G^+ \subset G$ (the "positive cone") such that $G^+ - G^+ = G$ and $G^+ \cap -G^+ = 0$. By an ideal of an ordered group is meant the subgroup generated by a face of the positive cone. F is a face of the positive cone of an ordered group if F is a subsemigroup and a hereditary subset (i.e., $0 \leq g \leq h$ with $g \in F$ implies $h \in F$). An ordered group is simple if it has no nonzero proper ideals. The quotient of an ordered group by an ideal is the quotient group ordered by the image of the positive cone. An ideal is prime if the quotient does not have two nonzero ideals with zero intersection.

The main results of sections 3 and 4 (there are also some results on ordered groups not satisfying the minimum condition on ideals — see 4.6 and 4.8) are obtained by using only a weaker property of dimension groups, the Riesz decomposition property, which states that the sum of two intervals is an interval. (The interval [0,a] of course is the set of all b with $0 \leq b \leq a$.) Ordered groups with this property were called Riesz groups by Fuchs in [13] (see also [17]). Clearly, a totally ordered group is a Riesz group, and also a direct sum or an inductive limit of Riesz groups is a Riesz group. It is slightly less trivial that also a lexicographical direct sum of Riesz groups is a Riesz group — see 3.10. It is not clear just which Riesz groups are dimension groups.

I thank H. Araki and N. Yui for discussions at various stages of the evolution of Theorem 2.2.

2. Ultrasimplicial order.

2.1. Definition. A subsemigroup of an abelian group will be said to be simplicial if it is generated by a finite independent set, and to be ultrasimplicial if it is the union of an upward directed collection of simplicial subsemigroups.

An ordered abelian group will be said to be simplicially ordered if its positive cone is simplicial, and to be ultrasimplicially ordered if its positive cone is ultrasimplicial.

Only abelian groups will be considered in this paper.

Clearly, a simplicially ordered group is just one which is isomorphic to the ordered group direct sum of finitely many copies of the ordered group $\mathbb{Z}$ (with positive cone $\mathbb{Z}^+ = \{0, 1, 2, \cdots\}$). A countable ultrasimplicially ordered group is just an ordered group isomorphic to the inductive limit of a sequence of simplicially ordered groups, with injective (positive) maps. In particular, such an ordered group is a dimension group (see 1), but the two classes of ordered groups are different (see 2.7).

An ultrasimplicially ordered group is easily seen to be torsion-free. Note that also a totally ordered group must be torsion-free.

<u>2.2. Theorem</u>. A totally ordered group is ultrasimplicially ordered.

<u>Proof</u>. Let G be a totally ordered group, and let S be a finite subset of G^+. We must find an independent subset B of G^+ such that the semigroup generated by B contains S.

We may suppose that the subgroup generated by S is equal to G (since a subgroup of a totally ordered group is totally ordered). Choose an arbitrary basis B_1 for the (finitely generated torsion-free commutative) group G. Replacing the negative elements of B_1 by their negatives, we may assume that $B_1 \subset G^+$.

By induction, we may suppose that all but one of the elements of S belongs to the semigroup generated by B_1, and then need only find a second basis $B_2 \subset G^+$ such that the semigroup generated by B_2 contains B_1 and also the single positive element g. Write

$$g = n_1 g_1 + \cdots + n_k g_k$$

where $B_1 = \{g_1, \cdots, g_k\}$. We shall construct B_2 by making a finite number of substitutions of g_i by $g_i - g_j$ where $g_i > g_j$, so that eventually all coefficients n_p are positive. (Then the altered B_1 generates a larger semigroup, also containing g, so it satisfies the requirements for B_2.)

First, let us show by induction on k that it is sufficient to consider the case that all but one of the coefficients n_p are positive. In other words, assume that B_2 exists for $k - 1$ in place of k, and let us reduce the problem for k to the case that only one n_p is negative. If there exist negative coefficients n_p and n_q with $p \neq q$, then $g - n_p g_p = \Sigma_{i \neq p} n_i g_i$ is positive (in fact $\geq g$), so by the inductive assumption may be rewritten to have all coefficients positive (by changing $\{g_i | i \neq p\}$ so as to generate a larger semigroup). Then we have $g = \Sigma n_i g_i$ with only n_p negative. Renumbering, we may suppose that only n_1 is negative.

Now $g = n_1 g_1 + n_2 g_2 + \cdots + n_k g_k$ with $n_1 \leq 0$ and $n_2, \cdots, n_k > 0$. If $n_1 = 0$, we are finished, so we may suppose that $n_1 < 0$. Suppose that for some $i = 2, \cdots, k$, $g_i > g_1$. Then replacing g_i by $g_i - g_1$, we have

$$g = (n_1 + n_i) g_1 + n_2 g_2 + \cdots + n_k g_k.$$

If $n_1 + n_i \geq 0$, we are finished. If $n_1 + n_i < 0$, then $|n_1 + n_i| = -n_1 - n_i = |n_1| - n_i < |n_1|$, so we have reduced the sum of the absolute values of the coefficients. Assuming inductively that B_2 exists for $\Sigma m_j g_j \in G^+$ whenever m_1 is the only negative coefficient and $\Sigma |m_j| < \Sigma |n_j|$ (for an arbitrary basis B_1), we may pass to the case that for all $i = 2, \cdots, k$, $g_1 > g_i$. Renumbering $g_2, \cdots, g_k$, we thus may suppose that

$$g_1 > g_2 > \cdots > g_k.$$

Since $\Sigma n_i g_1 > \Sigma n_i g_i > 0$, $\Sigma n_i > 0$. Denote by $r = 2, \cdots, k$ the index such that $n_1 + n_2 + \cdots + n_r \geq 0$ and $n_1 + \cdots + n_{r-1} < 0$. Consider the basis $B_1' = \{g_1', \cdots, g_k'\}$ where

$$g_1' = g_1 - g_r, \ldots, g_{r-1}' = g_{r-1} - g_r,\ g_r' = g_r, \ldots, g_k' = g_k.$$

Then the semigroup generated by B_1' contains B_1, and we have $g = \Sigma n_i' g_i'$ where

$$n_1' = n_1, \ldots, n_{r-1}' = n_{r-1}, n_r' = n_1 + \cdots + n_{r-1} + n_r,\ n_{r+1}' = n_{r+1}, \ldots, n_k' = n_k.$$

It is clear that $\Sigma |n_i'| < \Sigma |n_i|$, so by the inductive assumption of the preceding paragraph (which is the second in a double induction for the construction of B_2), B_2 exists as desired.

2.3. Corollary (stated in 5.2 of [10]). Every torsion-free group can be ultrasimplicially ordered.

Proof. Let G be a torsion-free (abelian) group. The group $G \otimes \mathbb{Q}$ has a basis over $\mathbb{Q}$, and can be lexicographically ordered with respect to this basis. The relative order on G is total, and therefore by 2.2 ultrasimplicial.

If the cardinality of G is at most that of $\mathbb{R}$, then G may be embedded as a subgroup of $\mathbb{R}$ and in this way given an ultrasimplicial order which is simple.

2.4. Corollary. The tensor product of two totally ordered groups is ultrasimplicially ordered, and if it is simple has a unique nonzero morphism into $\mathbb{R}$ (up to positive multiples). The nonzero positive elements in the tensor product of two simple totally ordered groups (which is simple) are precisely those elements with nonzero positive image in $\mathbb{R}$.

Proof. The tensor product of ultrasimplicially ordered groups is clearly ultrasimplicially ordered, so the first statement follows from 2.2.

Since the second statement holds for simple totally ordered groups, it is easily seen that any morphism from the tensor product of two simple totally ordered groups must be the tensor product of two morphisms (i.e., $\varphi\otimes\psi: g\otimes h \mapsto \varphi(g)\psi(h)$), and hence unique, up to a positive multiple.

Now let G and H be simple totally ordered groups, and $\varphi: G \to \mathbb{R}$, $\psi: H \to \mathbb{R}$ nonzero morphisms. Since G and H are simple, φ and ψ are injective on G^+ and H^+. In particular, if $(\varphi\otimes\psi)(\Sigma\, g_i\otimes h_i) = 0$ and all $g_i \in G^+$, $h_i \in H^+$, then $\Sigma\, g_i\otimes h_i = 0$. (It follows that $G\otimes H$ is simple.) Conversely, let us show that if $k \in G\otimes H$ is such that $(\varphi\otimes\psi)(k) > 0$, then $k > 0$, that is, $k = \Sigma\, g_i\otimes h_i$ with all $g_i \in G^+$, $h_i \in H^+$ (and of course not all $g_i\otimes h_i$ zero). We may assume G,H have finite rank.

Embed G and H in Euclidean spaces E and F so that rk G = dim E and rk H = dim F. Denote by $\overline{\varphi}$ and $\overline{\psi}$ the real-linear extensions of φ to E and ψ to F; then Ker $\overline{\varphi}$ and Ker $\overline{\psi}$ are the unique hyperplanes supporting G^+ in E and H^+ in F, respectively. If (S_n) (resp. (T_n)) is an increasing sequence of simplicial semigroups with union G^+ (resp. H^+), as given by 2.2, then for large n the hyperplanes supporting S_n (resp. T_n) are close to Ker $\overline{\varphi}$ (resp. Ker $\overline{\psi}$), so the hyperplanes supporting $S_n \otimes T_n$ are close to Ker $\overline{\varphi}\otimes\overline{\psi}$. In particular, if $k \in G \otimes H$ and $(\varphi\otimes\psi)(k)$ $(=(\overline{\varphi}\otimes\overline{\psi})(k)) > 0$ then for sufficiently large n, $k \in \mathbb{R}^+S_n \otimes \mathbb{R}^+T_n$. Since S_n and T_n are simplicial, so also is $S_n \otimes T_n$. It follows that if n is also large enough that $k \in S_n \otimes T_n - S_n \otimes T_n$, then $k \in S_n \otimes T_n$, and so $k \in (G \otimes H)^+$.

2.5. Lemma. Let G be an ordered group, and let H be an ideal of G. Suppose that both H^+ and $(G/H)^+$ are ultrasimplicial, and that the preimage in G of any nonzero element of $(G/H)^+$ is contained in G^+. Then G^+ is ultrasimplicial.

Proof. Let F be a finite subset of G^+. We must show that F is contained in a simplicial subsemigroup of G^+.

Choose a finite independent subset $\dot{S}$ of the ultrasimplicial semigroup $(G/H)^+$ such that the image of F in G/H is contained in the semigroup generated by $\dot{S}$. Choose a minimal subset S of G^+ with image $\dot{S}$; then S is independent.

Choose a finite subset F_H of H^+ such that F is contained in the group generated by $F_H \cup S$, and $F \cap H$ is contained in the semigroup generated by F_H. Since H^+ is ultrasimplicial we may choose F_H to be independent.

Then $F_H \cup S$ is independent. If for some $h \in F_H$ the h-coordinate of some $g \in F$ is not positive — in this case, $g \notin H$ —, this can be rectified by replacing some $k \in S$ such that the k-coordinate of g is nonzero by $k - nh$ where $-n$ is the h-coordinate of g. After finitely many such substitutions (of $k \in S$ by $k - nh$ with $h \in F_H$ and $n = 1, 2, \cdots$), F is contained in the semigroup generated by $F_H \cup S$.

2.6. Corollary. A lattice-ordered group with at most a finite number of independent ideals is ultrasimplically ordered.

Proof. By Theorem 1 of [4], such a lattice-ordered group is obtained from finitely many totally ordered groups by a finite number of steps of one of two kinds: passing to a finite direct sum, and passing to an extension in which the order is determined by the specified ideal and quotient as in 2.5 (if the ideal is nonzero, the quotient must of course be totally ordered for the extension to be lattice-ordered).

By 2.2, a totally ordered group is ultrasimplicially ordered. It is clear that the class of ultrasimplicially ordered groups is closed under direct sums, and by 2.5 it is closed under extensions of the kind under consideration.

By Theorem 6.1 of [5], an analogous decomposition exists also for a lattice-ordered group every singly generated ideal of which contains at most a finite number of independent ideals. Hence also such a lattice-ordered group is ultrasimplicially ordered.

2.7. Remark. Not every dimension group is ultrasimplicially ordered. The inductive limit of the sequence $\mathbb{Z}^3 \to \mathbb{Z}^3 \to \cdots$ in which each map is given by the matrix with positive entries

$$\begin{bmatrix} 1 & 1 & 1 \\ 2 & 1 & 0 \\ 0 & 1 & 2 \end{bmatrix}$$

is by definition a dimension group (see 1), but it is not ultrasimplicially ordered.

Indeed, calculation shows that the inductive limit is the group $\mathbb{Z}\left[\frac{1}{3}\right] \oplus \mathbb{Z}$ ($\mathbb{Z}\left[\frac{1}{3}\right]$ denotes the triadic rationals) with positive cone the set of elements with nonzero positive first coordinate, together with 0. Suppose that the positive elements $(1,-1)$, $(1,0)$ and $(1,1)$ were contained in the simplicial semigroup generated by the positive elements (a_1,b_1), (a_2,b_2) ; then $a_1,a_2 > 0$, and b_1 and b_2 are nonzero and have opposite signs. We have:

$$m_1(a_1,b_1) + m_2(a_2,b_2) = (1,0) \quad \text{for some} \quad m_1,m_2 \in \mathbb{Z}^+ ;$$

$$n_1(a_1,b_1) + n_2(a_2,b_2) = (1,1) \quad \text{for some} \quad n_1,n_2 \in \mathbb{Z}^+ .$$

From $(n_1-m_1)a_1 + (n_2-m_2)a_2 = 0$ it follows that $n_1 - m_1$ and $n_2 - m_2$ are either both zero or have opposite signs. Either possibility contradicts $(n_1-m_1)b_1 + (n_2-m_2)b_2 = 1$.

3. Lexicographical order.

3.1. Definition. An ordered group G will be said to be lexicographically ordered if an element g of G, with positive image in G/H whenever H is an ideal of G not containing g and maximal with this property, must be positive.

3.2. Lemma. Let G be a Riesz group. Then a finite intersection or sum of ideals of G is an ideal of G. The ideals of G form a distributive lattice. The quotient of G by an ideal is a Riesz group. The image of an ideal of G in the quotient by another ideal of G is an ideal of the quotient.

Proof. The first three statements are, respectively, 5.4 and 5.5, 5.6, and 5.3 of [13].

The last statement follows from the special case that the first ideal contains the second, together with the fact that the sum of two ideals of G is an ideal. To prove this special case, that if $I \supset J$ are ideals of G then I/J is an ideal of G/J, assume that $0 \leq a + J \leq b + J$ in G/J, and $b \in I$; then, replacing b by $a + c$ where $c \geq 0$ and $c + J = b - a + J$, we deduce that $a \in I$.

3.3 Lemma. Let G be a Riesz group, and let $H_1, \cdots, H_n$ be distinct maximal ideals of G with intersection 0. Then the canonical injection from G into $G/H_1 \oplus \cdots \oplus G/H_n$ is surjective, and is an order isomorphism.

Proof. For each $p = 1,\cdots,n$, denote by I_p the intersection of all H_q for $q \neq p$. Then for $q \neq p$ the image of I_p in G/H_q is 0. The image of I_p in G/H_p, which by 3.2 is an ideal of G/H_p, must be either 0 or all of G/H_p since H_p is maximal (that is, G/H_p is simple). To prove the lemma, then, it is enough to show that the image of I_p in G/H_p cannot be 0.

Using induction, we may suppose that the lemma holds for $n - 1$ maximal ideals with intersection 0. If the image of, say, I_n in G/H_n were 0, so that $I_n = I_n \cap H_n$, then $I_n = 0$; that is, $H_1 \cap \cdots \cap H_{n-1} = 0$. By the inductive assumption, G would then be isomorphic to $G/H_1 \oplus \cdots \oplus G/H_{n-1}$. This ordered group, however, has only $n - 1$ maximal ideals, in contradiction to the assumption that $H_1,\cdots,H_n$ are distinct. This shows that the image of I_p in G/H_p cannot be 0.

3.4. Lemma. Let G be a Riesz group such that each decreasing sequence $H_1 \supset H_2 \supset \cdots$ of ideals of G is finite. Then G has only finitely many maximal ideals.

Proof. Suppose that G has an infinite sequence $(I_1, I_2, \cdots)$ of distinct maximal ideals. We shall obtain a contradiction.

Set $I_1 = H_1$, $I_1 \cap I_2 = H_2$, $I_1 \cap I_2 \cap I_3 = H_3$, and so on. By hypothesis, for some $n = 1,2,\cdots$, $H_n = H_{n+p}$ for all $p = 1,2,\cdots$; in particular, $H_n = H_{n+1}$. Applying 3.3 to the Riesz group G/H_n and the distinct maximal ideals $I_1/H_n,\cdots,I_n/H_n$ (see 3.2), we obtain that G/H_n is isomorphic to $G/I_1 \oplus\cdots\oplus G/I_n$. (Recall that $(G/H_n)/(I_k/H_n)$ is isomorphic to G/I_k.) Similarly, G/H_{n+1} is isomorphic to $G/I_1 \oplus\cdots\oplus G/I_{n+1}$. Hence G/H_n has n distinct minimal ideals and G/H_{n+1} has $n + 1$; this contradicts $H_n = H_{n+1}$.

<u>3.5. Lemma</u>. Let G be a Riesz group such that each decreasing sequence $H_1 \supset H_2 \supset\cdots$ of ideals of G is finite. Then any intersection of ideals of G is an ideal.

<u>Proof</u>. Let S be a collection of ideals of G, and denote by I the intersection of these ideals. By 3.2 we may suppose that S is downward directed.

Choose $I_1 \in S$. If $I_1 \neq I$, there exists $I_2 \in S$ with $I_1 \supsetneqq I_2$. Similarly, if $I_2 \neq I$ there exists $I_3 \in S$ with $I_2 \supsetneqq I_3$. Thus, if $I \notin S$, and in particular if I is not an ideal, we may continue and obtain an infinite decreasing chain of ideals of G, in contradiction to the hypothesis on G.

<u>3.6. Theorem</u>. Let G be a Riesz group such that each simple subquotient of G is isomorphic to $\mathbb{Z}$. Assume that each decreasing sequence $H_1 \supset H_2 \supset\cdots$ of ideals of G is finite. Then G is lexicographically ordered.

<u>Proof</u>. Let $g \in G$ be such that the image of g in G/H is positive whenever H is an ideal of G not containing g and maximal with this property. We must show that g is positive.

By 3.5, there is a smallest ideal of G containing g. Since this ideal satisfies the same hypotheses as G, it is enough to consider the case that the ideal generated by g is equal to G. The assumption on g is then that it has positive image in each simple quotient of G.

By 3.4, G has only finitely many maximal ideals. By 3.2 (or 3.5), the intersection of the maximal ideals of G is an ideal; denote it by H.

For any $h \in H$, $g - h$ is contained in no maximal ideal of G, since g is contained in none and h in every one. In other words, $g - h$ has the same properties as g: it is positive in every simple quotient of G, and generates G as an ideal.

By 3.3, G/H is isomorphic to the direct sum of the simple quotients of G. In particular, the image of g in G/H is positive. Therefore, to show that g is positive, it is enough, replacing g by $g - h_0$ for a suitable $h_0 \in H$, to suppose that g itself is positive and prove that also $g - h$ is positive for every $h \in H$. Clearly, for this it is enough to consider the case that h is positive.

Let then $h \in H^+$; we must prove that $h \leq g$ (under the assumption that g is positive). Denote by I the ideal of G generated by h (see 3.5), and by J the intersection of the finitely many maximal ideals of I (see 3.4 and 3.2).

By transfinite induction on the ordered set of ideals of G — a technique applicable to any ordered set satisfying the decreasing chain condition (if such a set is totally ordered then it is well ordered, but the technique does not require total order) —, we may suppose that $h - k$ is positive for any $k \in J$.

It is enough to prove that $h + J \leq g + J$. Indeed, replacing h by $2h$ in this inequality, we have $2h - k \leq g$ for some $k \in J$. By the inductive hypothesis, $h - k$ is positive; hence $h \leq g$.

Passing to the quotient by J, then, we may suppose that $J = 0$ (see 3.2). Then by 3.3, I is isomorphic to the direct sum of its minimal ideals. Each of these is by hypothesis isomorphic to $\mathbb{Z}$. Therefore $h = h_1 + \cdots + h_n$ where each h_i is an atom in I^+.

Since g is positive (by provisional assumption), and the ideal generated by g is G, the face of G^+ generated by g is G^+. By the Riesz decomposition property, which says that a finite sum of intervals is an interval, the infinite sum of intervals $[0,g] + [0,g] + \cdots$ is a face, and therefore equal to G^+. In particular, $h_1 = g_1 + \cdots + g_m$

with $g_1,\cdots,g_m \in [0,g]$. Since I is an ideal, I^+ is a face and so $g_1,\cdots,g_m \in I^+$. By minimality of h_1, h_1 is equal to some g_i (and the others are 0), and so $h_1 \le g$.

Repeating the argument of the preceding paragraph with g replaced by $g - h_1$ (which is now known to be positive), we obtain $h_2 \le g - h_1$. Continuing, eventually we obtain $h_n \le g - (h_1+\cdots+h_{n-1})$, that is, $h \le g$.

3.7. Remark. A simple Riesz group is isomorphic to $\mathbb{Z}$ if it has a minimal (nonzero) positive element.

3.8. Problem. Does 3.6 still hold if the decreasing chain condition on ideals is dropped?

3.9. Corollary. Let G be a Riesz group such that each simple subquotient is isomorphic to $\mathbb{Z}$, and such that the spectrum of G is well ordered. Then G is totally ordered.

Proof. By 5.3, 10.1 and 10.2 of [13], every ideal of G is an intersection of prime ideals. By hypothesis, the prime ideals of G form a chain. Hence every ideal of G is the intersection of a chain of prime ideals, and is therefore prime.

By assumption, then, the set of all ideals of G is well ordered, and in particular satisfies the decreasing chain condition. Hence by 3.6, G is lexicographically ordered.

Let $g \in G$. Denote by H the largest ideal of G not containing g. (This exists as the ideals of G form a chain.) Denote by K the smallest ideal of G containing both H and g (see 3.5); then K/H is simple, and therefore by assumption isomorphic to $\mathbb{Z}$ and so totally ordered. So either $g + H$ or $-g + H$ belongs to $(G/H)^+$. Since G is lexicographically ordered (and H is the only ideal of G maximal not containing g), either $g \in G^+$ or $-g \in G^+$.

<u>3.10. Theorem</u>. The class of lexicographically ordered groups (resp. Riesz groups, dimension groups, ultrasimplicially ordered groups) is closed under the operation of forming lexicographical direct sums.

<u>Proof</u>. Let I be an ordered set, and for each $i \in I$ let G_i be an ordered group. Denote by G the lexicographical direct sum $\oplus_{i\in I} G_i$, that is, the group direct sum with a finite sum $\Sigma_{i\in F} g_i$ with $0 \neq g_i \in G_i$ for $i \in F$ being positive if (and only if) $g_i \in G_i^+$ for all maximal $i \in F$.

Suppose that each G_i is lexicographically ordered, and let $g \in G$ be such that $g + H \geq 0$ whenever H is an ideal of G maximal not containing H. Writing $g = \Sigma_{i\in F} g_i$ with $0 \neq g_i \in G_i$, to show that $g \geq 0$ we must show that $g_i \geq 0$ for i maximal in F. Fix i maximal in F, and let J be an ideal of G_i maximal not containing g_i. Then $J + \Sigma_{k \neq i} G_k = L$ is an ideal of G maximal not containing g_i Moreover, L contains g_k for $i \neq k \in F$, so it is maximal not containing g. By hypothesis, then, $g_i + L = g + L \geq 0$.

Equivalently (since $(G_i+L)/L$ is isomorphic to G_i/J), $g_i + J \geq 0$ in G_i/J. This shows that $g_i \geq 0$.

Suppose now that each G_i is a Riesz group. To show that G is a Riesz group, we may suppose that I is finite and use induction on the number of elements in I. Fix i maximal in I, and set $\Sigma_{k \neq i} G_k = H$. Then H is an ideal of G and by the inductive hypothesis, H is a Riesz group. Since $G = H \dot{+} G_i$, by Theorem 2.1 of [17] (see also 6.1 of [13]), to show that G is a Riesz group, that is, that

$$[0,g] + [0,h] = [0,g+h] \text{ for all } g,h \in G^+,$$

it is enough to prove this equality in the case that $g,h \in G_i^+$. Then $[0,g+h] \subset \Sigma_{k \leq i} G_k$. This shows that we may suppose that i is the largest element of I. Then $0 \neq b + H \geq 0$ implies $b \geq 0$, and so

$$[0,g] = H^+ \cup H+(]0,g[\cap G_i) \cup (g - H^+),$$

and similarly for h and $g + h$ in place of g. It is now clear that $[0,g] + [0,h] = [0,g+h]$ (since this holds inside G_i).

Suppose next that each G_i is a dimension group. To include the case of uncountable groups let us reformulate the definition of dimension group as follows (for G): for every finite subset $F \subset G^+$ there should exist for some $n = 1, 2, \cdots$ a positive map $\mathbb{Z}^n \to G$ and a subset F_0 of $(\mathbb{Z}^n)^+ (=(\mathbb{Z}^+)^n)$ with image F such that F_0 has the same rank in $\mathbb{Z}^n$ as F does in G. This done, we note that we may suppose that all G_i and the index set I are countable (that is, that G is countable). Then each G_i is the inductive limit of a sequence $G_i^{(1)} \rightarrow G_i^{(2)} \rightarrow \cdots$ of simplicially ordered groups, so that G is the inductive limit of the sequence $G^{(1)} \rightarrow G^{(2)} \rightarrow \cdots$ where $G^{(k)}$ is the lexicographical direct sum of the $G_i^{(k)}$, $i \in I$. It is sufficient to show that each $G^{(k)}$ is a dimension group; in other words, we may suppose that each G_i is simplicially ordered. Since we may suppose that I is finite, it follows from 4.9 (for example) that G is now in fact ultrasimplicially ordered. (This is also easily verified directly.)

Suppose finally that each G_i is ultrasimplicially ordered. Then the sequences $G_i^{(1)} \rightarrow G_i^{(2)} \rightarrow \cdots$ of the preceding paragraph can be chosen with injective maps, so G as given is ultrasimplicially ordered.

4. Skeletons.

4.1. Definition (cf. [16]). The skeleton of a Riesz group G is the set of all pairs (π, G_π) where π is a prime ideal of G such that the quotient ordered group G/π has a nonzero simple ideal G_π (necessarily unique — see 3.2). Denote the set of such prime ideals of G by $S(G)$.

One may consider the Jacobson (hull-kernel) topology on $S(G)$, but here we shall consider just the order given by inclusion. (The order in $S(G)$ gives the closures of points in the Jacobson topology. If $S(G)$ is totally ordered, or if the set of all finite intersections of ideals $\pi \in S(G)$ satisfies the decreasing chain condition, every closed set in $S(G)$ is a finite union of point closures.)

The skeleton of a Riesz group G can be used to define another Riesz group with the same skeleton — the lexicographical direct sum $\oplus_{\pi \in S(G)} G_\pi$ of the simple ideals of prime quotients of G (see 3.10). If, then, G is determined by its skeleton, it is isomorphic to a lexicographical direct sum of simple ordered groups, and in particular is lexicographically ordered (see 3.10). (If, moreover, each simple subquotient of G is ultrasimplicially ordered, G is then ultrasimplicially ordered — see 3.10; cf. 4.9.)

4.2. Theorem. Let G be a lexicographically ordered Riesz group. Suppose that each simple subquotient of G lifts. (In other words, if G/H is simple then $0 \to H \to G \to G/H \to 0$ splits, with positive maps.)

Suppose that every decreasing sequence $H_1 \supset H_2 \supset \dots$ of ideals of G is finite. Then G is determined by its skeleton.

Proof. By transfinite induction we may suppose that G has a maximal ideal H and that H is determined by its skeleton.

There exists a smallest ideal of G not contained in H. Indeed, the intersection of two ideals of G not contained in H is an ideal by 3.2, and is not contained in H since H is maximal (so by 3.2, if $H_1 \not\subset H$, $H_2 \not\subset H$ then $H + H_1 \cap H_2 = (H+H_1) \cap (H+H_2) = G$; that is, $H_1 \cap H_2 \not\subset H$). This shows that the collection of all ideals of G not contained in H is downward directed, and hence as in 3.5 has a smallest member. Denote this smallest ideal of G not contained in H by I.

Suppose that $g \in I$, $g \notin H$, and $g + H \geq 0$. Let J be an ideal of G which is maximal such that $g \notin J$. Then $J \cap I \neq I$, so $J \cap I \subset H$. Since $(J+H) \cap I = J \cap I + H \cap I$ (by distributivity — see 3.2), it follows that $g \notin J + H$, and therefore $J + H = J$, so $J \supset H$ and hence $J = H$. In particular, $g + J \geq 0$. Since G is lexicographically ordered, $g \geq 0$. Since by assumption $I = H \cap I \dotplus K$ with K an ordered group lifting of $I/(H \cap I)$, and $H \cap I$ is determined by its skeleton, this shows that I is determined by its skeleton.

It follows from the fact that H and I are determined by their skeletons that G is determined by its skeleton. We have $G = H \dotplus K$ (since $G = H + I$ and $I = H \cap I \dotplus K$), and

H is isomorphic to $\oplus_{\pi\in S(H)} H_\pi$, the lexicographical direct sum of the simple subquotients of H. Thus, identifying $S(G)$ with $S(H) \cup \{H\}$, we have that G is isomorphic as a group to $\oplus_{\pi\in S(G)} G_\pi$ where $G_\pi = H_\pi$ for $\pi \in S(H)$ and $G_H = K$. Suppose that for a finite set $F \subset S(G)$, $0 \neq g_\pi \in G_\pi$ for $\pi \in F$, and $g_\pi \geq 0$ if π is maximal in F. We need only consider the case that $H \in F$. We have $\Sigma g_\pi = \Sigma_{\pi\subset H} g_\pi + \Sigma_{\pi\not\subset H} g_\pi$. The first sum is positive because $\pi \in H$ implies $\pi \not\supset I$ and $g_\pi \in I_\pi$, and the second because $\pi \not\subset H$ implies $\pi \neq H$ and $g_\pi \in H_\pi$.

4.3. Remark. In 4.2, the condition that each simple subquotient lift is clearly necessary for the conclusion to hold (see also 4.7). The decreasing chain condition on ideals is of course not necessary (see also 4.6), but cannot just be dropped (see 4.8).

4.4. Corollary. Let G be a Riesz group such that each simple subquotient of G is isomorphic to $\mathbb{Z}$ and such that each decreasing sequence $H_1 \supset H_2 \supset \cdots$ of ideals of G is finite. Then G is determined by its skeleton.

Proof. By 3.6, G is lexicographically ordered. Apply 4.2.

4.5. Corollary. Let G be a lexicographically ordered Riesz group such that each simple subquotient of G is divisible and such that each decreasing sequence $H_1 \supset H_2 \supset \cdots$ of ideals of G is finite. Then G is determined by its skeleton.

Proof. To apply 4.2, we need only show that each simple subquotient of G lifts. By transfinite induction, G is divisible, so certainly each simple subquotient of G lifts as a group.

Passing to an ideal of G, we must show that if H is a maximal ideal of G then there exists a subgroup K of G such that the canonical map $G \to G/H$ restricted to K is an order isomorphism of K onto G/H. Passing to the smallest ideal of G not contained in H, as in the proof of 4.2, we may suppose that H is the largest proper ideal of G. Then it is sufficient just to choose a group lifting of G/H — any subgroup K with $G = H \dot{+} K$ is automatically an ordered group lifting. This follows from the argument in the third paragraph of the proof of 4.2. (For any $g \notin H$, $g + H \geq 0$ implies $g \geq 0$.)

4.6. Theorem. Let G be a countable divisible totally ordered group. Then G is determined by its skeleton.

Proof. The proof is very similar to the proof of J. Erdös in [12] that a totally ordered vector space of countable dimension over the real numbers is determined by its skeleton. Here, the reals are replaced by the rationals, and an extra step is needed because the simple subquotients need not be of dimension one over the rational field $\mathbb{Q}$. Since the proof is more elaborate because of this extra step, it is given here in full.

Choose a basis $(a_1,a_2,\cdots)$ for the $\mathbb{Q}$-vector space G. We may choose $(a_1,a_2,\cdots)$ such that if certain a_i each generate the same ideal of G then their images in the nonzero simple quotient of this ideal are independent. (If a_n generates the same ideal as certain a_i with $i < n$, and if the images of a_n and these a_i in the nonzero simple quotient of this ideal are dependent, while the images of the a_i alone are independent, just change a_n by subtracting a suitable linear combination of these a_i so that a_n generates a smaller ideal. After finitely many such changes of a_n, this situation no longer arises. Carry out this procedure for each $n = 2,3,\cdots$ to obtain a basis with the desired property.)

Now group together those a_i generating the same ideal. If I is a singly generated ideal of G, then it is generated by some a_i. Denote by G_I the $\mathbb{Q}$-vector space spanned by those a_i generating the ideal I; G_I is isomorphic to the nonzero simple quotient of I. Indeed, by the choice of the basis $(a_1,a_2,\cdots)$, the canonical map of I to its simple quotient is injective on G_I. Also, any element of I is a linear combination of certain a_i, which by the choice of the basis must lie in I, and in fact must lie either in G_I or in the maximal proper ideal of I. This shows that any element of I belongs to G_I modulo the maximal proper ideal of I.

It follows that G is the direct sum of subgroups lifting the simple subquotients of G. Since G is totally ordered, it follows immediately that G is isomorphic to a lexicographical direct sum of simple totally ordered groups; in other words, G is determined by its skeleton.

4.7. Example. Totally ordered groups of rank two are not in general determined by their skeletons. Indeed, any torsion-free commutative group of rank two can be given a total order in which it is not simple, and if, as a group, it is not the direct sum of two subgroups of rank one, then as an ordered group it is certainly not the lexicographical direct sum of its simple subquotients.

4.8. Example. Totally ordered groups with simple subquotients isomorphic to $\mathbb{Z}$ are not in general determined by their skeletons. While the lexicographical direct sum $\oplus_{n \in \mathbb{Z}} \mathbb{Z}$, say G_1, is isomorphic to the lexicographical direct sum of its simple subquotients, this is not true for the subgroup G_2 of G_1 consisting of those elements with even coordinate sum. Indeed, G_2 has the same skeleton as G_1, but is not isomorphic to G_1 (as an ordered group — of course as a group G_1 is isomorphic to G_2). If an element of G_1 is divisible by 2 in every quotient by a nonzero ideal, then all its coordinates and hence the element itself are divisible by 2. The element $(\cdots,0,2,0,\cdots)$ of G_2, however, while divisible by 2 in any quotient of G_2 by a nonzero ideal (which is equal to the quotient of G_1 by the corresponding ideal of G_1), is not divisible by 2 in G_2 because the element $(\cdots,0,1,0,\cdots)$ of G_1 has odd coordinate sum.

Replacing 2 by $3,4,\cdots$, one obtains a whole sequence $(G_n)_{n=1,2,\cdots}$ of ordered groups with the same skeleton, no two of which are isomorphic. Indeed, if $n_1 \neq n_2$ then for some prime p, n_1 and n_2 have different powers of p as factors, so for some $q = 1,2,\cdots$, p^q divides both n_1 and n_2, and p^{q+1} divides n_1 (say) but not n_2. It follows that if an element of G_{n_2} is divisible by p^q modulo each nonzero ideal of G_{n_2}, then it is divisible by p in G_{n_2}; G_{n_1} on the other hand does not have this property.

In a similar way it is also possible to construct an uncountable family of pairwise nonisomorphic sub ordered groups of G_1 having the same skeleton as G_1. (This condition on the skeleton is satisfied by any subgroup separating the ideals of G_1.) One family of such subgroups is obtained by choosing a fixed sequence $(S_1, S_2, \cdots)$ of pairwise disjoint infinite subsets of the index set $\mathbb{Z}$, and to any sequence $(p_1, p_2, \cdots)$ of prime numbers associating the subgroup $G_{(p_k)}$ of all elements of G_1 such that for each $k = 1,2,\cdots$ the sum of the coordinates over the indices in S_k is divisible by p_k. While it does not seem immediately clear in general that two distinct sequences of primes determine nonisomorphic ordered groups, this is clear in the case that the two sequences involve different sets of primes. (In this case, for some prime p, one subgroup contains an element which is divisible by p modulo every nonzero ideal but not in the subgroup itself, while the other subgroup, which does not have p appearing in its defining sequence of primes, does not have such an element.) Since there are uncountably many sets of primes, there are uncountably many isomorphism classes of ordered groups obtained by this construction.

4.9. The skeleton is also useful in cases where it does not determine the ordered group. Aside from the Hahn embedding theorem for totally ordered groups (see also [4]), one has for example the following result.

Theorem. Let G be a lexicographically ordered Riesz group. Suppose that each simple subquotient of G is ultrasimplicially ordered. Assume that each decreasing sequence $H_1 \supset H_2 \supset \cdots$ of ideals of G is finite. Then G is ultrasimplicially ordered.

Proof. It follows from the hypotheses that $na \geq 0$ for some $n = 2,3,\cdots$ implies $a \geq 0$. (If H is an ideal of G maximal such that $a \notin H$ then by the decreasing chain condition $a + H$ is contained in a simple ideal of G/H, ultrasimplicially ordered by hypothesis, so $na + H \geq 0$ implies $a + H \geq 0$. Since G is lexicographically ordered, $a \geq 0$.)

This says that $G \subset G \otimes \mathbb{Q}$ and $G^+ = (G \otimes \mathbb{Q})^+ \cap G$. Since $G \otimes \mathbb{Q}$ satisfies the hypotheses of 4.5 (it is a Riesz group by 2.5 of [13]), it is determined by its skeleton.

By induction we may suppose that G has a maximal ideal H and that H is ultrasimplicially ordered. We may also suppose, provided that we extend this to G, that a simplicial subsemigroup of H^+ containing a given finite subset of H^+ can be chosen to be generated by an independent set of elements of the simple groups in some decomposition of $H \otimes \mathbb{Q}$ as the lexicographical direct sum of simple ordered groups.

Let F be a finite subset of G^+. Then as in 2.5 choose $F_H \subset H^+$ and $S \subset G^+ \setminus H$ such that: $F_H \cup S$ is independent, the subsemigroup generated by F_H contains $F \cap H$, the subgroup generated by $F_H \cup S$ contains F, and, in G/H, the image of S is independent and the subsemigroup it generates contains the image of F.

Denote by J the smallest ideal of G not contained in H — see 4.2. Since H is maximal, $G = J + H$, and we may choose S to lie in J^+. Choose a subgroup K of $J \otimes \mathbb{Q}$ containing S such that $G \otimes \mathbb{Q} = H \otimes \mathbb{Q} \dotplus K$ (recall that S is independent modulo H). As in 4.2, such a subgroup K is necessarily an ordered group lifting of the quotient (cf. 4.5). Choosing F_H as permitted by the inductive assumption, we then have that both F_H and S consist of elements of the simple groups in a decomposition of $G \otimes \mathbb{Q}$ as a lexicographical direct sum of simple ordered groups.

If the integral linear combination of elements of $F_H \cup S$ giving some element of F has a negative nonzero coefficient, the position of this coefficient cannot be maximal among the nonzero coefficients, so as in 2.5 we may replace one of the maximal elements of $F_H \cup S$ by a smaller element of G to make this coefficient positive. This change will not make any other coefficients negative in the expressions of elements of F with respect to $F_H \cup S$. Therefore if such a change is performed sufficiently many times, all the coefficients become positive.

It must be verified that the new $F_H \cup S$ satisfies the inductive hypothesis that for a suitable decomposition of $G \otimes \mathbb{Q}$ as a lexicographical direct sum of simple ordered groups, each element of $F_H \cup S$ belongs to one of the simple subgroups. This holds for the originally chosen $F_H \cup S$. It is enough to consider one step in the changing of $F_H \cup S$, in which $g \in (G\otimes\mathbb{Q})_\pi$, say, is replaced by $g - h$ with $h \in (G\otimes\mathbb{Q})_{\pi'}$, for some $\pi' \leq \pi$. But inside the ideal generated by $(G\otimes\mathbb{Q})_\pi$, any group lifting of the unique nonzero simple quotient is an ordered group lifting; in particular, we may choose a group lifting containing the element $g - h$ of this ideal, together with the other elements of $F_H \cup S$ originally contained in $(G\otimes\mathbb{Q})_\pi$ (which together with $g - h$ form an independent set).

<u>4.10. Corollary</u>. Let G be a Riesz group such that each simple subquotient of G is totally ordered, and suppose that each decreasing sequence $H_1 \supset H_2 \supset \cdots$ of ideals of G is finite. Then G is ultrasimplicially ordered.

<u>Proof</u>. (In the rather special case that G is a lattice-ordered group, this corollary is a consequence of 2.6.)

To show that G is lexicographically ordered, suppose that $g + H \geq 0$ whenever H is an ideal of G maximal not containing g. If $g \not\geq 0$ then by 10.2 of [13] there is a prime ideal I of G such that $g + I \not\geq 0$. Then with $H \supset I$ an ideal of G maximal not containing g (such an ideal of G exists by Zorn's lemma), $g + H < 0$, in contradiction to the supposition on g. It follows that $g \geq 0$.

By 2.2, each simple subquotient of G is ultrasimplicially ordered.

Hence by 4.9, G is ultrasimplicially ordered.

5. Applications to C*-algebras.

5.1. Classification of approximately finite-dimensional separable C*-algebras.

Let us recall the classification theorem for inductive limits of sequences of finite-dimensional C*-algebras given in [10]. Such C*-algebras (which are separable) we shall call approximately finite-dimensional — for short, AF-algebras.

Let A be an AF-algebra. The (abstract) dimension on A is defined to be the map d from projections in A to their Murray-von Neumann equivalence classes. Thus, if e is a projection in A (i.e., $e^2 = e = e^*$), then d(e) denotes the set of projections $\{uu^* \mid u \in A, u^*u = e\}$. Denote the range of d — the range of the dimension on A — by D(A). The operation of adding two orthogonal projections in A induces a (partially defined) binary operation in D(A), making D(A) a commutative local semigroup. Thus, d has the two properties $d(u^*u) = d(uu^*)$ and $d(e+f) = d(e) + d(f)$; moreover, d is determined by these properties — every such map on projections of A factors through d.

The fundamental fact concerning the local semigroup D(A) is that it is a complete invariant for A. In other words, for any two AF-algebras A and A', the local semigroups D(A) and D(A') are isomorphic if and only if A and A' are isomorphic. In fact, more is true: D is a covariant functor, and any isomorphism of D(A) and D(A') is induced by an isomorphism of A and A'.

The functor D preserves finite direct sums and preserves inductive limits of sequences. If M_n denotes the C*-algebra of $n \times n$ complex matrices then $D(M_n) = \{0,1,\cdots,n\}$. Since any finite-dimensional C*-algebra is isomorphic to a direct sum of various M_n, these properties may be used to compute D(A) for any AF-algebra A.

If A is an AF-algebra then D(A) may be embedded in a unique way as a generating subset of a group (which is equal to the group $K_0(A)$ of algebraic K-theory). This group is torsion-free, and moreover the semigroup generated by D(A) has zero intersection with its negative; so with this semigroup as positive cone the group becomes an ordered group which we shall denote by G(A), and call the dimension group of A. The ordered group G(A) is the inductive limit of a sequence of direct sums of copies of $\mathbb{Z}$ — a dimension group in the sense of 1 —, and is an arbitrary such ordered group. As a group, G(A) is an arbitrary countable torsion-free group (see 2.3). The dimension range, D(A), is an upward directed, hereditary (that is, $x \in D(A)$ and $0 \leq y \leq x$ imply $y \in D(A)$), generating subset of the positive cone of the dimension group G(A), and every such subset of a dimension group is isomorphic to the range of the dimension on some AF-algebra.

Closed two-sided ideals of an AF-algebra are in bijective correspondence with their images in G(A), and also with the subgroups of G(A) generated by these images, which are arbitrary ideals of the ordered group $G(A)$ (that is, positively generated subgroups closed under taking intervals). A hereditary sub-C*-algebra of an AF-algebra is also an AF-algebra (by 3.1 of [11]) and has the same dimension group as the closed two-sided ideal it generates.

The functor G, as well as preserving inductive limits of sequences, preserves tensor products; this is obvious for finite-dimensional C*-algebras, and if $A = \varinjlim A_i$ and $B = \varinjlim B_i$ with A_i and B_i finite-dimensional, then

$$\begin{aligned} G(A\otimes B) &= G(\varinjlim A_i \otimes \varinjlim B_i) = G(\varinjlim (A_i \otimes B_i)) \\ &= \varinjlim G(A_i \otimes B_i) = \varinjlim (G(A_i)\otimes G(B_i)) \\ &= \varinjlim G(A_i) \otimes \varinjlim G(B_i) = G(A) \otimes G(B). \end{aligned}$$

The functor G also preserves short exact sequences. Again, this is obvious for finite-dimensional C*-algebras. Let

$$0 \to A \to B \to C \to 0$$

be a short exact sequence of AF-algebras, with $B = \varinjlim B_i$, B_i finite-dimensional. Then by 3.1 of [2], $A = \varinjlim A_i$ where A_i is the preimage in A of B_i, and of course $C = \varinjlim C_i$ where C_i is the image in C of B_i. For each i the short sequence

$$0 \to A_i \to B_i \to C_i \to 0$$

is exact, and the inductive limit of this sequence of short exact sequences is the given short exact sequence. Since G preserves inductive limits, the short sequence

$$0 \to G(A) \to G(B) \to G(C) \to 0$$

is the inductive limit of the sequence of short exact sequences

$$0 \to G(A_i) \to G(B_i) \to G(C_i) \to 0,$$

and is therefore exact (as a little diagram chasing shows).

Important examples of AF-algebras are the UHF-algebras, or, what it seems natural to call the Glimm algebras, M_n, where n is a generalized integer, that is, an infinite product of prime powers $2^a 3^b 5^c 7^d \cdots$ where $a, b, \cdots = 1, 2, \cdots$ or ∞ (see [14] and [8]); $D(M_n)$ is the set of all rational numbers in the interval $[0,1]$ whose denominators divide n. Also important is the elementary C*-algebra M_∞; $D(M_\infty) = \mathbb{Z}^+$.

5.2. Postliminary C*-algebras with minimum condition on ideals.

By means of 4.4 and the invariant described in 5.1, it is possible to rederive the classification of separable postliminary C*-algebras with finitely many closed two-sided ideals given by Behncke and Leptin in [1], and, at the same time, to extend this classification to the class of separable postliminary C*-algebras with the decreasing chain condition on closed two-sided ideals (or, equivalently, the condition on the spectrum that, when considered as an ordered set, every subset have at least one minimal element, and at most finitely many). One obtains in particular that there is a unique such algebra with spectrum order isomorphic to $\mathbb{N}$ — this answers a question of Behncke and Leptin (see lines

3-5, page 256 of [1]).

The statement is as follows. Let A be a separable postliminary C*-algebra such that each decreasing sequence of closed two-sided ideals of A is finite. Then A is approximately finite-dimensional, and the dimension group G(A) of A (see 5.1) is isomorphic to the lexicographical direct sum

$$\oplus_{t\in A^\wedge} \mathbb{Z}$$

where $A^\wedge$ is the spectrum of A, identified with the primitive spectrum of A and ordered by inclusion. (It follows that the Jacobson topology on $A^\wedge$ is determined by the order structure.) There exists a smallest closed two-sided ideal I of A such that A/I has a unit. (For A/I to be nonzero it is of course necessary that $A^\wedge$ have a maximal element.) The range of the dimension on A, a complete invariant for A (see 5.1), is determined by the ordered set $A^\wedge$, together with the final subset $(A/I)^\wedge$, and the positive element of the lexicographical direct sum $\oplus_{t\in(A/I)^\wedge} \mathbb{Z}$ which corresponds to the image of the unit of A/I in G(A/I) by an identification of these two ordered groups. Since this identification of $\oplus_{t\in(A/I)^\wedge} \mathbb{Z}$ with G(A/I) is not unique, to give this last piece of information in an invariant way we must rather specify the set of all the positive elements which correspond to the image of the unit of A/I in G(A/I) by different identifications of $\oplus_{t\in(A/I)^\wedge} \mathbb{Z}$ with G(A/I). This set is an orbit under the group of order automorphisms of $\oplus_{t\in(A/I)^\wedge}\mathbb{Z}$ preserving ideals, which consists of those maps changing each coordinate by adding an integral combination of coordinates in higher positions. For t maximal in $(A/I)^\wedge$, the coordinate at t is of course uniquely determined — it is the order of the simple finite-dimensional C*-algebra A/t.

Let us first show that A must be an AF-algebra. By transfinite induction, it is enough to consider the case that there is a closed two-sided ideal I of A such that I is an AF-algebra and A/I is simple, and therefore isomorphic to the AF-algebra M_n for some $n = 1,2,\cdots,\infty$. In this case it follows by 7 of [3] (which depends on the assumption that I is postliminary) that A is an AF-algebra.

Since the ideal structure of $G(A)$ is the same as that of A, and every prime closed two-sided ideal of A is primitive and therefore contained in a smallest larger ideal, with quotient isomorphic to M_∞, so that the skeleton of $G(A)$ is just $\{(t,\mathbb{Z}) \mid t \in A^\wedge\}$, it follows by 4.4 that $G(A)$ is isomorphic to the lexicographical direct sum $\oplus_{t\in A^\wedge}\mathbb{Z}$.

To show that there exists a smallest closed two-sided ideal of A modulo which A has a unit, it is enough to show that if I and J are closed two-sided ideals such that A/I and A/J have units, then also $A/(I\cap J)$ has a unit. This is easiest to prove using that A is an AF-algebra; in this case there are projections e,f in A which are units modulo I and J respectively. (There is a finite-dimensional sub-C*-algebra B of A such that the image of

B in A/I contains an element strictly within distance 1 of the unit of A/I, and therefore also contains the unit of A/I; this is then the image of a projection e in B.) There is then a single projection g in A such that $\|ge-e\| < 1$, $\|gf-f\| < 1$. (The unit g of a finite-dimensional sub-C*-algebra of A containing elements x,y with $\|e-x\| < \frac{1}{2}$, $\|f-y\| < \frac{1}{2}$ satisfies: $gx = x$; $\|ge-e\| \leq \|ge-gx\| + \|gx-x\| + \|x-e\| < \frac{1}{2} + \frac{1}{2} = 1$; also $\|gf-f\| < 1$.) These conditions imply that g is a unit modulo both I and J — equivalently, modulo $I \cap J$.

Note that no nonzero quotient of I has a unit; if J is a closed two-sided ideal of I such that I/J has a unit then I/J is a direct summand of A/J and so $A/J = A/I \oplus I/J$ has a unit, whence by minimality of I, $J = I$.

To show that the dimension range is determined by the invariants mentioned (which in the case $I = A$ reduce to just the ordered set $A^\wedge$), we must show that $D(A)$ is the preimage of $D(A/I)$ in $G(A)^+$. Since $D(A)$ is a hereditary subset of $G(A)^+$, for this it is enough to show that $D(A) + G(I)^+ = D(A)$; indeed, with $g \in D(A)$ and $h \in G(I)^+$, it is sufficient to find $k \in D(A)$ such that $k \geq g + h$. By 3.10, $G(A)$ is lexicographically ordered, so to show that $k \geq g + h$ it is sufficient to show that $k - (g+h)$ is positive modulo each ideal maximal not containing it — or, since $k \geq 0$, just modulo each ideal maximal not containing $g + h$. By 3.5, there is a smallest ideal containing $g + h$, and by 3.4 this has only finitely many maximal ideals, so there are only finitely many ideals of $G(A)$ maximal not containing $g + h$. Since

$D(A)$ is upward directed, it is enough to consider a single ideal H maximal not containing $g + h$, and find $k \geq g + h \pmod{H}$.

Consider first the case that H is not a maximal ideal in $G(A)$. Then there is a proper ideal K of $G(A)$ strictly containing H, and therefore also containing $g + h$. Since $D(A)$ generates $G(A)$ we may choose $k \in D(A)$ with $k \notin K$. Then $k \geq g + h \pmod{H}$.

Now consider the case that H is a maximal ideal in $G(A)$, so that $G(A)/H = \mathbb{Z}$. If $h \in H$, then $g \geq g + h \pmod{H}$ so we may take $k = g$. If $h \notin H$, so that $G(I) \not\subset H$, then $(D(A))+H)/H$ does not have a maximal element (as $G(I)$ is the smallest ideal modulo which $D(A)$ has a maximal element), so $(D(A)+H)/H = \mathbb{Z}^{+}$, and there exists $k \geq g + h \pmod{H}$ (such k exists, in fact, for arbitrary g and h in $G(A)$).

It remains to determine the order automorphisms of the lexicographical direct sum $\oplus_{t \in (A/I)^{\wedge}} \mathbb{Z} = G$ which preserve ideals. First, any map $\varphi: G \to G$ changing each coordinate by adding certain integral multiples (depending on φ) of coordinates in higher positions (these, of course, add up to a single integral multiple of the greatest common divisor of the coordinates in higher positions) is clearly additive. Since φ leaves fixed all nonzero coordinates of an element which are maximal in position among the nonzero coordinates, φ is injective, and takes each positive element into a positive element generating the same ideal. To show that φ is surjective, by transfinite induction it is sufficient to prove this assuming that the restriction of φ to H is surjective with H a maximal ideal of G. Since φ induces e identity in G/H (it fixes each coordinate in maximal sition, in particular the coordinate which is zero for l members of H), it is then clear that φ is surjective. is shows that φ is an ideal-preserving order automorphism G.

Conversely, let φ be an ideal-preserving automorphism the ordered group G. To show that φ is of the form nsidered in the preceding paragraph, it is enough to show at it agrees on a generating set with such an automorphism. king the canonical basis of G, we see that it is enough show that φ fixes each nonzero coordinate of an element ich is maximal in position among the nonzero coordinates. ssing to a prime quotient, it is enough to show that φ is e identity on the smallest nonzero ideal of this quotient; is ideal is isomorphic to $\mathbb{Z}$, an ordered group with only e automorphism.

3. Some antiliminary C*-algebras with postliminary quotients.

The totally ordered groups G_n, $n = 1,2,\cdots$, described 4.8 give rise by 2.2 and 5.1 to a sequence of pairwise nisomorphic antiliminary AF-algebras, with primitive ectrum order isomorphic to $\{-\infty\} \cup \mathbb{Z}$, and with all proper otients isomorphic to the unique separable C*-algebra ose spectrum is order isomorphic to $\mathbb{N}$ (see 5.2). The nstruction in 4.8 of an uncountable family of pairwise nisomorphic totally ordered groups with the same skeleton the lexicographical direct sum $G_1 = \oplus_{n\in\mathbb{Z}} \mathbb{Z}$ shows that there are in fact uncountably many C*-algebras with these properties (in contrast with 5.2, where at most countably many C*-algebras, and often just one, correspond to each of the skeletons considered).

By 4.6 and 5.1, the tensor products of any two AF-algebras with the above properties and with totally ordered dimension groups by the largest Glimm algebra $M_{2^\infty 3^\infty 5^\infty 7^\infty \cdots}$ are isomorphic.

Shifting coordinates by one position to the left is an automorphism of the lexicographical direct sum $G_1 = \oplus_{n\in\mathbb{Z}} \mathbb{Z}$, leaving invariant for each $n = 1,2,\cdots$ the subgroup G_n of elements with coordinate sum divisible by n. By 5.1, for each $n = 1,2,\cdots$ the automorphism of G_n defined in this way is induced by an automorphism of the (unique) AF-algebra $A(G_n)$ with dimension range G_n^+. While the algebra $A(G_n)$ itself is unique, up to isomorphism (uniqueness holds even if just the dimension group is specified, since G_n^+ has no proper generating hereditary subsets and so the dimension range must be all of G_n^+ — cf. 5.2), there are at first sight many automorphisms of $A(G_n)$ inducing the shift in G_n, just as there are many automorphisms of $A(G_n)$ inducing the identity in G_n (namely, the inner automorphisms and their simple limits). It is possible, however, that any two automorphisms of $A(G_n)$ inducing the shift in G_n determine the same crossed product C*-algebra.

The crossed product of $A(G_1)$ by one particular isomorphism inducing the shift in G_1 has been studied by Cuntz — if, as in [7], O_∞ denotes the C*-algebra generated by an infinite sequence of isometries with orthogonal ranges (Cuntz showed that this is unique), then $O_\infty \otimes M_\infty$ is the crossed product of $A(G_1)$ by the automorphism described as follows. Consider the map $x \mapsto 1 \otimes x \otimes e$ of the C*-algebra $\tilde{M}_\infty$ into $\tilde{M}_\infty \otimes \tilde{M}_\infty \otimes \tilde{M}_\infty$ where $\tilde{M}_\infty$ is M_∞ with unit adjoined and e is a fixed minimal projection of M_∞, and the analogous map $x \mapsto 1 \otimes x \otimes e$ of $\otimes_{i=1,2,3}\tilde{M}_\infty$ into $\otimes_{i=1,\cdots,5}\tilde{M}_\infty$, and so on; then the C*-algebra $A(G_1)$ is isomorphic to a closed two-sided ideal of the inductive limit of the sequence $(\otimes_{i=1,\cdots,2k+1}\tilde{M}_\infty)$ — the sub-C*-algebra generated by the sequence $(\otimes_{i=1,\cdots,2k+1}M_\infty)$. The endomorphisms $1 \otimes x \mapsto x \otimes e$ of $\otimes_{i=1,\cdots,2k+1}\tilde{M}_\infty$ extend one another and determine an automorphism of the inductive limit, leaving invariant the ideal isomorphic to $A(G_1)$. This automorphism of $A(G_1)$ induces the shift in G_1; indeed, it is a shift itself (it is the restriction of the shift to the left in the infinite tensor product over $\mathbb{Z}$ of copies of $\tilde{M}_\infty$, tailing off to the left as 1 and to the right as e).

A somewhat puzzling point is the following: while it is easy to construct a sub-C*-algebra of $A(G_1)$ isomorphic to $A(G_n)$, $n = 2,\cdots$, it does not seem clear how to do this in a shift-invariant way.

5.4. Simple AF-algebras with comparability of projections.

Let A be an AF-algebra. Suppose that any two projections in A are comparable in the sense of Murray and von Neumann, that is, that one of them is the support of a partial isometry with range contained in the other. Then G(A), the dimension group of A (see 5.1) is totally ordered. To see this, note that the assumption states that D(A), the dimension range of A (see 5.1), is totally ordered; this can be restated as

$$D(A) - D(A) \subset D(A) \cup -D(A).$$

Since the semigroup generated by D(A) is $G(A)^+$, it follows by induction that

$$G(A) = G(A)^+ - G(A)^+ \subset G(A)^+ \cup -G(A)^+,$$

in other words, that G(A) is totally ordered.

Conversely, if G(A) is totally ordered, then so also is the subset D(A) of G(A) — that is, projections in A are comparable.

Since closed two-sided ideals of A correspond precisely to ideals of G(A) (that is, to subgroups generated by faces of $G(A)^+$), A is simple if and only if G(A) is simple.

If G(A) is totally ordered then it is simple if and only if it is Archimedean (that is, $0 \leq na \leq b$ for all $n = 1,2,\cdots$ implies $a = 0$). Hence by [15], in this case G(A) is isomorphic to a sub ordered group of $\mathbb{R}$.

Now let G be a countable sub ordered group of $\mathbb{R}$ containing $\mathbb{Z}$. By 2.2 and 5.1, $G \cap [0,1]$ is isomorphic to the dimension range of an AF-algebra with unit.

Thus we see that the isomorphism classes of simple unital AF-algebras in which any two projections are comparable are in bijective correspondence with the countable subgroups of $\mathbb{R}$ containing $\mathbb{Z}$.

Tensoring with M_∞ introduces a degeneracy, as the range of the dimension becomes the whole positive cone of the dimension group. The resulting "properly infinite" AF-algebras are classified by the order isomorphism classes of countable subgroups of $\mathbb{R}$.

By 2.4 and 5.1 the class of simple AF-algebras with comparability of projections is not closed under tensor products. There is, however, by 2.2, an AF-algebra in this class associated naturally with the tensor product of two AF-algebras A and B in this class — the AF-algebra whose dimension range is the image of $D(A\otimes B)$ under the unique morphism $G(A\otimes B) \to \mathbb{R}$ (see 2.4 and 5.1).

The fact that a countable totally ordered group $G \subset \mathbb{R}$ is ultrasimplicially ordered (2.2), which is stronger than its being a dimension group (see 2.7), can be used to construct a concrete realization of the simple AF-algebra $A(G)$ whose dimension range is isomorphic to G^+. In the Hilbert space $\ell^2(G)$ consider the projections corresponding to the half-open intervals $[a,b[$ in G, and for each such projection e denote by $t(e)$ the length of the corresponding nterval. Consider also the partial isometries in $\ell^2(G)$ orresponding to the (partial) translations of one half-open nterval in G onto another. The C*-algebra B(G) generated y these partial isometries is simple by the Corollary to heorem 1 of [9]. It is easy to see that two projections ,f in C(G) are equivalent in B(G) if and only if .(e) = t(f). If G is of rank one, then the algebra generated y the partial translations connecting projections in C(G) s locally finite-dimensional, so B(G) is an AF-algebra nd every projection in B(G) is equivalent to one in :(G), whence the dimension range of B(G) is G^+. If G s not of rank one, then the algebra generated by the partial ranslations is not locally finite-dimensional, so B(G) s presumably not an AF-algebra. It is conceivable that very projection in B(G) is still equivalent to one in :(G), so that G^+ is always the dimension range of B(G); his would answer the question raised by Douglas in [9] hether B(G) determines the ordered group G.

It is at any rate possible to construct a sub-C*-algebra $C(G) \subset A \subset B(G)$ such that two projections e,f in C(G) are equivalent in A if and only if t(e) = t(f), and such that any projection in A is equivalent to a projection in C(G) — and, moreover, such that A is an AF-algebra, and therefore isomorphic to A(G). To show this it is enough to show that if M is a finite-dimensional sub involutive algebra of B(G) generated by partial translations, and if e and f are two projections in C(G) equivalent in B(G) then there exists a finite-dimensional sub involutive algebra N generated by partial translations such that N contains M and e and f, and e is equivalent to f in N. Let then such M and e and f be given. Since the algebra obtained by adjoining a projection in $C(G)$ to M is finite-dimensional, we may suppose that e and f are in M. Now choose minimal projections $p_1,\cdots,p_k$ in M, one from each minimal direct summand of M. By 2.2 there exists an independent subset $\{g_1,\cdots,g_r\}$ of G^+ such that the semigroup generated by $g_1,\cdots,g_r$ contains the elements $t(p_1),\cdots,t(p_k)$ of G^+. It follows that each p_i may be partitioned into projections of the form $q \in C(G)$ with $t(q) \in \{g_1,\cdots,g_r\}$; choose such a decomposition of each of $p_1,\cdots,p_k$. For each $i = 1,\cdots,r$ fix one of the projections q with $t(q) = g_i$ in the decomposition of one of $p_1,\cdots,p_k$, and choose partial isometries consisting of sums of partial translations connecting this projection to the other projections in the decompositions of $p_1,\cdots,p_k$ to which it is equivalent in $B(G)$. The algebra generated by M and these partial isometries is finite-dimensional; set it equal to N. N has r minimal direct summands $N_1,\cdots,N_r$, and for each $i = 1,\cdots,r$, if q is a minimal projection in N_i then $t(q) = g_i$. Since $t(e) = t(f)$ and $g_1,\cdots,g_r$ are independent, it follows that e is equivalent to f in N (in fact any two projections of N are equivalent in N if they are equivalent in $B(G)$).

While a sub-C*-algebra A of $B(G)$ with the properties specified at the beginning of the preceding paragraph is unique up to isomorphism, it is not actually unique, unless G is isomorphic to $\mathbb{Z}$. (Since a single partial translation always generates a finite-dimensional sub-C*-algebra of $B(G)$, uniqueness of A would imply that A must contain every partial translation and so be equal to $B(G)$. This is probably impossible if G is not of rank one; in any case, if A is constructed (as above) as the closure of a locally finite-dimensional involutive algebra generated by partial translations, it is easy to see that A cannot be equal to $B(G)$ if rank $G \neq 1$, and that A is not necessarily equal to $B(G)$ if rank $G = 1$ and G is not isomorphic to $\mathbb{Z}$.) Thus, one has not obtained a functor inverse to D.

Nevertheless, at least for $G \subset \mathbb{Q}$ there is a canonical choice of an AF-algebra with dimension range isomorphic to G^+, namely, $B(G)$. If $G = \mathbb{Z}[1/n]$, $n = 1,2,\cdots$ (the group of n-adic rationals), then multiplication by n is an order automorphism of G; the crossed product of $B(G)$ by the automorphism induced by this automorphism of G is just $O_n \otimes M_\infty$ where O_n is the C*-algebra with unit generated by n isometries with orthogonal ranges with sum 1, shown by Cuntz in [7] to be unique. It would be interesting to study the automorphisms of $B(G)$ induced by order automorphisms of G for other subgroups G of $\mathbb{Q}$, or of $\mathbb{R}$.

If a countable subgroup G of $\mathbb{R}$ has an order automorphism α, by 5.1 α is induced by an automorphism of the AF-algebra with dimension range isomorphic to G^+. If G

is not of rank one, it does not seem to be possible to represent such an automorphism as the restriction to a subalgebra of the automorphism of B(G) induced by α. This problem is analogous to that referred to in the last paragraph of 5.3.

R E F E R E N C E S

1. H. Behncke and H. Leptin, Classification of C*-algebras with a finite dual, J. Functional Analysis 16 (1974), 241-257.

2. O. Bratteli, Inductive limits of finite dimensional C*-algebras, Trans. Amer. Math. Soc. 171 (1972), 195-234.

3. O. Bratteli and G.A. Elliott, Structure spaces of approximately finite-dimensional C*-algebras II, J. Functional Analysis (to appear).

4. P. Conrad, The structure of a lattice-ordered group with a finite number of disjoint elements, Michigan Math. J. 7 (1960), 171-180.

5. P. Conrad, Some structure theorems for lattice-ordered groups, Trans. Amer. Math. Soc. 99 (1961), 212-240.

6. P. Conrad, J. Harvey and C. Holland, The Hahn embedding theorem for abelian lattice-ordered groups, Trans. Amer. Math. Soc. 108 (1963), 143-169.

7. J. Cuntz, Simple C*-algebras generated by isometries, Comm. Math. Phys. 57 (1977), 173-185.

8. J. Dixmier, On some C*-algebras considered by Glimm, J. Functional Analysis 1 (1967), 182-203.

9. R.G. Douglas, On the C*-algebra of a one-parameter semigroup of isometries, Acta Math. 128 (1972), 143-151.

10. G.A. Elliott, On the classification of inductive limits of sequences of semisimple finite-dimensional algebras, J. Algebra 38 (1976), 29-44.

11. G.A. Elliott, Automorphisms determined by multipliers on ideals of a C*-algebra, J. Functional Analysis 23 (1976), 1-10.

12. J. Erdös, On the structure of ordered real vector spaces, Publ. Math. Debrecen 4 (1956), 334-343.

13. L. Fuchs, Riesz groups, Ann. Scuola Norm. Pisa III 19 (1965), 1-34.

14. J. Glimm, On a certain class of operator algebras, Trans. Amer. Math. Soc. 95 (1960), 318-340.

15. O. Hölder, Die Axiome der Quantität und die Lehre vom Mass, Ber. Verh. Sächs. Akad. Wiss. Leipzig, Math. Phys. Cl. 53 (1901), 1-64.

16. P. Ribenboim, Sur les groupes totalement ordonnés et l'arithmétique des anneaux de valuation, Summa Brasil. Math. 4 (1958), 1-64.

17. J.R. Teller, On partially ordered groups satisfying the Riesz interpolation property, Proc. Amer. Math. Soc. 16 (1965), 1392-1400.

SEMIPRIME CROSSED PRODUCTS

Joe W. Fisher
Department of Mathematical Sciences
University of Cincinnati
Cincinnati, Ohio 45221

During the last year the question of when the crossed product of any group ove a semiprime coefficient ring is itself semiprime has been settled by Montgomery, Passman, and the author in [1], [2], and [3]. I would like to bring the readers attention to this body of work by presenting the main results and by contrasting both the results and the techniques to the analoguous ones for group rings. Since the amount of material in these papers is large and the exposition is excellent, I must set very limited objectives for this brief account with the hope that the interested reader will consult those papers. In the interests of the casual reader I will present the story as it unfolded during the last year rather than just givi the ending of the story in its most elegant polished form. In order to keep thing technically simple, I will present everything in the context of skew group rings e though all the results stated for skew groups rings hold verbatim for crossed prod I would like to thank the organizers, John Lawrence and David Handelman for inviti me to participate in this conference.

Let R be a ring with 1 and let G be a multiplicative group. By the crossed pr duct $R\dagger G$ of R with G we mean an associative ring which is determined by R, G, and certain other parameters. More precisely $R\dagger G$ is the free left R-module on $\{\bar{g} : g \in$ with multiplication given by the formulas $r\bar{g} = \bar{g}r^{g}$ and $\bar{g}\bar{h} = t(g,h)\overline{gh}$ for all g, h r in R. Here $t : G \times G \to U$ is a map from $G \times G$ to the group of units of R and fixed g in G the map $\bar{g} : r \to r^{\bar{g}}$ is an automorphism of R. It is a simple matter determine the relations on t and the automorphisms $\{\bar{g} : g \in G\}$ in order to antee that R†G is associative. Moreover, R†G has a unity element $1 = t(1,1)^{-1}\bar{e}$ h we may assume without loss of generality is $1 = \bar{e}$. Also $\mathcal{H} = \{u\bar{g} : u \in U, \bar{g} \in G$ multiplicative group of units in R†G. Thus the equation $r\bar{g} = \bar{g}r^{\bar{g}}$ is equivalent $r^{\bar{g}} = \bar{g}^{-1}r\bar{g}$ and hence the automorphism $\bar{g}$ is merely conjugation by a unit in R†G. fact, it is clear that $\mathcal{H}$ acts on R by conjugation. In general, R†G does not tain an isomorphic copy of G; however, $U \triangleleft \mathcal{H}$ and $G \cong \mathcal{H}/U$.

If there is trivial skewing, i.e., $r^{\bar{g}} = r$ for all r in R and g in G, then R†G ices to what is called a <u>twisted group ring</u> denoted $R^t[G]$. Here it is clear that t(g,h) must belong to the center of R. If there is trivial twisting, i.e., 1, then we get what is called a <u>skew group ring</u> denoted R*G. If there is both ial skewing and trivial twisting, we get the ordinary <u>group ring</u> denoted R[G]. oth of these latter cases since $\bar{g}\bar{h} = \overline{gh}$, R†G contains a copy of G by setting g and we identify G as a subgroup of units of R†G.

The question we are interested in is the following.

<u>Question</u>. When is the skew group ring R*G semiprime?

This question initially came up in a conversation with Robert L. Snider at the 6 University of Oklahoma Ring Theory Symposium when we were discussing a possible ution to a problem which the author posed at that Symposium. Even though Farkas Snider [4] later solved the problem by using other techniques, the question of n is R*G semiprime remained.

We begin by considering the case when G is finite. First, it is obvious that in order for R*G to be semiprime, R must be semiprime. For if N is a nilpotent ideal of R, then $\sum_{g\in G} N^g$ is a nilpotent G-invariant ideal of R and $(\sum_{g\in G} N^g)*G$ is a nilpotent ideal of R*G. Second, if R is semiprime with additive $|G|$-torsion, then R*G need not necessarily be semiprime as the following example shows.

<u>Example 1.</u> (R. L. Snider) Let F be a field of characteristic p for which there exists a group G acting on F with $p \mid |G|$. Then G acts on $R = F \oplus F$ via $(f_1,f_2)^g = (f_1^g,f_2)$. Then $R*G \cong (F*G) \oplus F[G]$. Evidently, F[G] is not semiprime and hence R*G is not semiprime.

The question boils down to this. As in the case of ordinary group rings, is R semiprime with no $|G|$-torsion sufficient to guarantee that R*G is semiprime? The first result in this direction was obtained by Handelman-Lawrence-Schelter [5] in 1976 when they were able to prove it in the special case of G finite abelian. In 1977 Montgomery and I proved it in general for any finite group G. I want to sketch our proof, but first I need to develop the necessary machinery of Kharchenko [6].

Assume that R is semiprime and let $\mathcal{F}$ denote the filter of all essential two-sided ideals of R. Let $R_{\mathcal{F}}$ be the ring of quotients of R with respect to $\mathcal{F}$, i.e.,

$$R_{\mathcal{F}} = \varinjlim_{I\in\mathcal{F}} \mathrm{Hom}_R({}_RI,R).$$

This was defined for prime rings by W. S. Martindale [7] and extended to semiprime rings by S. A. Amitsur [8]. The elements of $R_{\mathcal{F}}$ are equivalence classes of left R-module homomorphisms from essential ideals of R into R. For each x in $R_{\mathcal{F}}$, let I he essential ideal of R associated with x, so that $0 \neq I_x x \subsetneq R$, when $x \neq 0$.

Then R may be embedded in R_F which is semiprime, and when R is prime, R_F is prime. If C denotes the center of R_F, then C is a von Neumann regular ring and when R is prime, C is a field. Each g in G may be extended to an automor-m of R_F as follows: if x is in R_F, then for a in I_x^g, let $a \cdot x^g = (a^{g^{-1}} x)^g$. x^g determines a left R-module homomorphism of I_x^g to R and thus x^g is a unique ent of R_F.

Define $\Phi_g = \{x \in R_F : xr^g = rx \text{ for all } r \in R\}$. It can be shown that $\{x \in R_F : xr^g = rx \text{ for all } r \in R_F\}$. Hence $\Phi_e = C$.

Definition. (Kharchenko [6]). If G is a group of automorphisms of R, set $= \{g \in G : \Phi_g \neq 0\}$.

The following example shows that G_{inn} is not necessarily a subgroup of G; however, s closed under taking inverses [1].

Example 2. Let F be a field and let R be a direct sum of 3 copies of F. Then S_3 s on R by permuting the summands. Here $R = R_F$ and it is easy to see that G_{inn} is a subgroup of S_3.

When R is prime, G_{inn} is always a subgroup of G, as was pointed out in [6, §2]. his case, nonzero elements of Φ_g are invertible in R_F, and so G_{inn} consists pre-ly of those elements of G which become inner (given by conjugation by a unit) extended to R_F.

The algebra of the group G is defined to be $B = \sum_{g\varepsilon G}\Phi_g$. Clearly B is a C-sub-algebra of R_F. Moreover, B is invariant under G since $\Phi_g^h = \Phi_{h^{-1}gh}$. Thus the skew group ring B*G is defined.

The reason for introducing R_F is that it is close to R, but yet large enough to contain certain additional units. The first lemma from [2] shows that R_F contains the units which we will need.

Lemma 3. Let σ be an automorphism of R and let $a,b \in R$ be fixed nonzero elements. If $arb = br^\sigma a^\sigma$ for all $r \in R$, then there exists a unit $x_\sigma \in \Phi_\sigma$ such that $b = ax_\sigma$.

Proof. Let $A = RaR$ and $B = RbR$ and define maps $f : A \to B$ and $h : B \to A$ by $f : \sum_i x_i a y_i \to \sum_i x_i b y_i^\sigma$ and $h : \sum_i x_i b y_i \to \sum_i x_i a y_i^{\sigma^{-1}}$. In order to see that f is well-defined, it suffices to show that $0 = \sum_i x_i a y_i$ implies that $0 = \sum_i x_i b y_i^\sigma$. To this end, suppose that $0 = \sum_i x_i a y_i$. Then for all $r \in R$, the formula $atb = bt^\sigma a^\sigma$ yields that $0 = (\sum x_i a y_i) rb = (\sum x_i b y_i^\sigma) r^\sigma a^\sigma$ and hence $0 = \sum x_i b y_i^\sigma$ since $a^\sigma \neq 0$ and R is prime. Similarly, if $0 = \sum_i x_i b y_i$, then for all $r \in R$, we have that

$0 = (\sum x_i b y_i) r a^\sigma = (\sum x_i a y_i^{\sigma^{-1}}) r^{\sigma^{-1}} b$. Hence $0 = \sum x_i a y_i^{\sigma^{-1}}$. Thus both f and h are well-defined and since they are clearly left R-homomorphisms, we have that $f = x_\sigma \in R_F$ and $h \in R_F$. Furthermore, $fh = 1$ on A and $hf = 1$ on B so $h = x_\sigma^{-1}$ and x_σ is a unit in R_F.

Observe that $a_r f$ is defined on R and for all $x \in R$ we have that $x(a_r f) = (xa)f = xb = xb_r$. Thus $ax_\sigma = b$. Finally, let $c \in R$. Then $hc_r f$ is defined on B and for all $xby \in B$ we have that $(xby)(hc_r f) = (xay^{\sigma^{-1}})(c_r f) = (xay^{\sigma^{-1}} c)f = xbyc^\sigma = (xby)c_r^\sigma$. Thus $x_\sigma^{-1} c x_\sigma = c^\sigma$ and $x_\sigma \in \Phi_\sigma$.

Now we are ready to prove that if G is finite and R is semiprime with no -torsion, then R*G is semiprime. The idea of the proof is simple! As we l see in a moment the problem can be reduced to the case where R is prime. n the problem of R*G being semiprime can be reduced to the problem of $B{*}G_{inn}$ ng semiprime. However, $B{*}G_{inn}$ is very nice. Indeed, it turns out to be isimple Artinian.

<u>Lemma 4</u>. Let R be a prime ring. If $B{*}G_{inn}$ is semiprime, then R*G is semi- me.

<u>Proof</u>. If R*G is not semiprime, then choose an ideal $N \neq 0$ of R*G with $N^2 = 0$. $\alpha = ae + \sum_{g\neq e} a_g g$ be a nonzero element of N of minimal support length. Then each $r \in R$, $ar\alpha - \alpha ra$ is an element of N of shorter length. Therefore, $\alpha = \alpha ra$. Let $g \in \mathrm{Supp}(\alpha)$. Now $(ara_g)g = ar(a_g g) = (a_g g)ra = (a_g r^{g^{-1}} a^{g^{-1}})g$. So $a_g = a_g r^{g^{-1}} a^{g^{-1}}$. By Lemma 3 there exists $x_{g^{-1}} \in \phi_{g^{-1}}$ such that $a_g = ax_{g^{-1}}$. Con- uently, $\alpha = a(\sum x_{g^{-1}} g)$ where each $x_{g^{-1}} \in \phi_{g^{-1}}$. Thus $0 \neq z = \sum x_{g^{-1}} g \in B{*}G_{inn}$.

Now from $\alpha(R{*}G)\alpha = 0$ one can show that $z(B{*}G_{inn})z = 0$. This is a bit technical l hinges on R being prime. The net result is that $B{*}G_{inn}$ is not semiprime. That the proof.

<u>Lemma 5</u>. If R is prime with no $|G|$-torsion, then $B{*}G_{inn}$ is semiprime.

<u>Proof</u>. For convenience, write $H = G_{inn}$. For each $g \in H$, we have that $\Phi_g = Cx_g$ some $x_g \in \Phi_g$. Moreover $x_g x_h = t(g,h)x_{gh}$ where $T : H \times H \to C$ since $\phi_g\phi_h \subseteq \phi_{gh}$. rthermore, it is easy to check that α is a factor set. Hence $B = \sum_{g\in H}\phi_g$ is a

homomorphic image of $C^t[H]$, the twisted group algebra of H over C with respect to t. By a version of Maschke's theorem for twisted group algebras, $C^t[H]$ is semisimple Artinian since C has no $|H|$-torsion. Whence B is semisimple Artinian. Now a version of Maschke's theorem for skew group rings applied to yield that $B*H$ is semisimple Artinian.

Theorem 6 (Fisher-Montgomery [1]) Let R be a semiprime ring and G a finite group acting as automorphisms on R. If R has no additive $|G|$-torsion, then $R*G$ is semiprime.

Proof. From the previous two lemmas we have the result for R prime. We need only to push it up to R semiprime.

The proof will proceed by induction on $|G|$. Since R has no $|G|$-torsion, $(0) = P_\alpha$ where the P_α's are prime ideals of R such that R/P_α has no $|G|$-torsion. For each P_α, let $I_\alpha = \bigcap_G P_\alpha^g$. Then R/I_α is semiprime with no $|G|$-torsion and has an induced action of G. Since $R*G$ is a subdirect product of the $(R/I_\alpha)*G$, it suffices to show that each $(R/I_\alpha)*G$ is semiprime.

If P_α is G-invariant, then $I_\alpha = P_\alpha$ and R/P_α is prime. Wherefore $(R/I_\alpha)*G$ is semiprime by Corollary 5. We may therefore assume that R contains a nonzero prime ideal P with $\bigcap_G P^g = (0)$.

Set orb $P = \{P^g : g \in G\}$ and let m be the smallest integer such that for any choice of m idstinct members of orb P, say $P_1,\ldots,P_m$, we have that $P_1 \cap P_2 \cap \ldots \cap P_m = (0)$. Clearly $1 < m \leq |G|$. Choose $P_1,\ldots,P_{m-1}$ in orb P such that $V = P_1 \cap \ldots \cap P_{m-1} \neq (0)$. Let $H = \{g \in G : g \text{ permutes } P_1,\ldots,P_{m-1}\}$. Then $|H| < |G|$ since G is transitive on orb P. Moreover, H acts on V and V has no $|H|$-torsion. Hence, by our induction hypothesis, $V*H$ is semiprime.

Note that if $g \notin H$, then $VgV = gV^gV = (0)$. Thus since V is H-invariant *H)g(V*H) = (0) for all $g \notin H$.

We claim that R*G is semiprime. For, if not, let N be an ideal of R*G with = (0). Applying Lemma 6, we may choose $x = \sum r_g g \in N$ with $r_g \in V$ for all $g \in G$ d $r_e \neq 0$. Write

$$x = \sum_{h \in H} r_h h + \sum_{g \notin H} r_g g = x_H + x_{G-H} \quad .$$

nce $N^2 = (0)$, we have that $(V*H)x(V*H)x(V*H) = (0)$. Using the fact that *H)x_{G-H}(V*H) = (0), one can show that $(V*H)x_H(V*H)x_H(V*H) = (0)$. But then x_H nerates a nilpotent ideal of V*H. A contradiction since $x_H \neq 0$. Therefore R*G semiprime and the proof is complete.

<u>Corollary 7</u>. ([3]) If R is semiprimitive with no $|G|$-torsion, then R*G is miprimitive.

<u>Proof.</u> The $\{g : g \in G\}$ form a normalizing basis for R*G over R. As is well-own $J(R*G)^{|G|} \subseteq J(R)(R*G) = 0$ [9, Theorem 7.2.5]. However, R*G is semiprime by eorem 6. Hence J(R*G) = 0.

The picture is complete for finite groups. But what about infinite groups? at can one expect there? To give some idea of what one might expect or hope for t us recall the result and technique for ordinary group rings. The result is e following.

<u>Theorem 8</u>. (Connell-Passman [10]) If R is semiprime with no $|N|$-torsion for ch finite normal subgroup N of G, then R[G] is semiprime.

The technique used is the so-called Δ-reduction technique which goes as follows. Let $\Delta = \Delta(G) = \{g \in G : g$ has only finitely many conjugates in $G\}$ and $\theta : R[G] \to R[\Delta]$ is the projection map, i.e., $\theta(\sum_{g\in G} r_g g) = \sum_{g\in\Delta} r_g g$. If A is a nonzero ideal in R[G], then $\theta(A)$ is a nonzero ideal of $R[\Delta]$. With the stated hypotheses the idea of the proof is (1) to show that $R[\Delta]$ is semiprime and (2) to show that if A is a nilpotent ideal of R[G], then $\theta(A)$ is a nilpotent ideal of $R[\Delta]$.

We immediately have (1) for skew group rings.

Theorem 9. (Fisher-Montgomery [1]) If R is semiprime with no $|N|$-torsion for each finite normal subgroup N of G, then $R*\Delta$ is semiprime.

Proof. A sequence of reductions similar to those used in the proof for ordinary group rings [10, p. 163] can be made to reduce the problem down to R*H wher H is finite so that Theorem 6 can be applied.

This leaves (2) to deal with for skew group rings. In order to handle (2) for ordinary group rings one proves that if $\alpha x\beta = \gamma x\delta$ for fixed α, β, γ, δ in R[G] and for all x in G, then $\theta(\alpha)\theta(\beta) = \theta(\gamma)\theta(\delta)$. The proof of this is rather involved. From this, of course, it follows that if A and B are ideals of R[G] with AB = 0, then $\theta(A)\theta(B) = 0$.

Initially, it appeared that the application of this technique to skew group rings was hopeless because of the complication introduced by the skewing of the coefficients when one moves them past group elements. Even though they obtained much less than originally hoped for, Montgomery and Passman [2, §1] did succeed with the Δ-reductions; however, they had to pay the price for the additional compl cation by assuming that R was prime. On the assumption that R is prime, they prov

t if $\alpha x\beta = \gamma x\delta$ for fixed α, β, γ, δ in R*G and for all x in G, then for <u>cer-</u>
<u>n</u> y in G only $y^{-1}\theta(\alpha)y\theta(\beta) = y^{-1}\theta(\gamma)y\theta(\delta)$. However, this was just enough to ow them to prove that if R is prime, then AB = 0 for ideals A and B of R*G im- es that $\theta(A)\theta(B) = 0$. Moreover, they produce an example to show that even if s prime, the conjugating element y is necessary in the above.

By making use of their Δ-reduction techniques they obtain many beautiful orems concerning the primeness and semiprimeness of R*G in the case when R is me. We list several of their results without proof since it would be too big a k to reproduce them here.

<u>Theorem 10</u>. (Montgomery-Passman [2]) Assume that R is prime. Then R*G is iprime if and only if for all finite normal subgroups N of G with $N \subseteq G_{inn}$, it ults that R*N is semiprime.

This theorem coupled with Theorem 6 shows that if R is prime of characteristic then R*G is semiprime. At characteristic p, their main theorem is the following. all that $\Delta^+(G) = \{x \in \Delta : x \text{ has finite order}\}$.

<u>Theorem 11</u>. (Montgomery-Passman [2]) Let R be a prime ring of characteristic > 0.

(i) If $\Delta^+(G) \cap G_{inn}$ contains no elements of order p, then R*G is semiprime.

(ii) R*G is semiprime if and only if R*P is semiprime for all finite elementary abelian p-subgroups P of $\Delta^+(G) \cap G_{inn}$.

By combining Theorem 6 and Theorem 10 we do obtain for skew groups rings with <u>ime coefficient rings</u> an analogue of the result for group rings, viz., if R is prime with no $|N|$-torsion for each finite normal subgroup N of G, then R*G is semiprime. What about the case when the coefficient ring is semiprime? In particular, if R is semiprime, do we get an analogue of the group ring result? The following example shows that we do not.

Example 12. (Passman [3]) Let G be an infinite simple group which contains a subgroup P of prime order p. Let Ω denote the set of right cosets of P in G. Then G acts transitively on Ω via right multiplication and P is the stabilizer of point $m \in \Omega$. Let K be a field of characteristic p. Define $R = \Pi_{\omega\in\Omega} K_{\omega}$ to be the complete direct product of copies of K, one for each $\omega \in \Omega$. Then R is a commutat regular ring with 1 and $pR = 0$.

Define an action of G on R by letting G permute the factors of R in precisely the same way that it permutes Ω. Let e_m be the idempotent in R which has 1 in th m-coordinate and 0 elsewhere. Then form the skew group ring R*G and let $f = \sum_{g\varepsilon P} g \in R*G$. A short calculation shows that $e_m f(R*G)e_m f = 0$ and hence $e_m f$ generates a nonzero ideal of square zero. Thence R*G is not semiprime.

Consequently, this example momentarily left the situation up in the air for t case of semiprime coefficient rings. And I say momentarily for it did not take Passman long to sift out a theorem which, of course, was not as nice as originall hoped for - however, it is indeed very nice in light of the above example. I wil state it now and try to give some idea of how the proof works. First some definitions.

Recall that G acts on R by conjugation and hence G permutes the ideals of R. If L is an annihilator ideal of R, then we let $G_L = \{g \in G : L^g = L\}$ be the stabilizer of L in G. Then G_L is a subgroup of G and we get an induced action of G_L o R/L. Let $G_{(L)}$ denote the subgroup of G_L which stabilizes all the annihilator als of R/L. It is clear that $G_{(L)}$ is a normal subgroup of G_L.

Theorem 13. (Passman [3]) Let R be semiprime. If R*G is not semiprime, then re is an annihilator ideal L of R and a finite normal subgroup $N \trianglelefteq G_{(L)} \trianglelefteq G_L$ h that R/L and hence R has $|N|$-torsion.

A few remarks are in order. First, if R is prime in the above, then L = 0 so) $= G_L = G$ and we conclude that G has a finite normal subgroup N such that R no $|N|$-torsion. Second, it is quite possible that the conclusion of the above orem can be strengthened to the statement that N is in fact normal in G_L (this true for R Noetherian). Indeed, - as Passman points out - it can be shown rly easily that G_L must have nontrivial finite normal subgroups. The problem t remains is to connect the orders of these subgroups to the torsion of R/L. rd, an immediate corollary of the Theorem which is easy to state is that if R semiprime and R has no $|x|$-torsion for each x of finite order in G, then R*G semiprime.

The idea of the proof is as follows. If R*G is not semiprime, then there exists onzero ideal I of R*G of square zero. Again let γ be a nonzero element of I with imal support length and e in the Supp γ. Say, Supp $\gamma = \{x_1 = e, x_2, \ldots, x_n\}$. ine $B_i = \{r \in R : \text{there exists } \beta = \sum_{j=1}^{n} r_j x_j \in I \text{ with } r_i = r\}$. Set $A = B_1$, ann(A) , and H = <Supp γ> . Then it can be shown that $H \subsetneq G_{(L)}$. Hence) $[R*G_{(L)}]$ is a nonzero ideal of $R*G_{(L)}$ of square zero. Moreover, the image $I \cap [R*G_{(L)}]$ gives a nonzero ideal of square zero in $(R/L)*G_{(L)}$. Therefore, the of has been reduced to the situation where R/L is semiprime, $(R/L)*G_{(L)}$ is not iprime, and by definition $G_{(L)}$ stabilizes all the annihilator ideals of R/L. Now guts of the proof is Passman's [3, Lemma 6] which states that if R is semiprime h no $|N|$-torsion for each finite normal subgroup N of G and if G acts on R so as to stabilize all the annihilator ideals of R, then R*G is semiprime. This Lemma comes from a careful blend of Theorem 9 together with the Δ-reduction technique i [2, Lemma 1.5]. From this it follows that $G_{(L)}$ contains a finite normal subgroup N such that R/L has |N|-torsion and hence R has |N|-torsion. So that is the idea of the proof.

As I mentioned in the beginning all the results so far hold for crossed products. A natural question is what happens in the special case of twisted group rings? Here the situation is nicer and one obtains an analogue of the group ring result.

Theorem 14. (Reid [11]) If R is semiprime with no |N|-torsion for each fini normal subgroup N of G, then the twisted group ring $R^t[G]$ is semiprime.

Of course this follows immediately from Passman's [3, Lemma 6] mentioned abov

References

J. W. Fisher and S. Montgomery, Semiprime skew group rings, J. Algebra, 52 (1978), 241-247.

S. Montgomery and D. S. Passman, Crossed products over prime rings, Israel J. Math., (to appear).

D. S. Passman, Crossed products over semiprime rings, (to appear).

D. R. Farkas and R. L. Snider, Noetherian fixed rings, Pacific J. Math., 69 (1977), 347-353.

D. Handelman, J. Lawrence and W. Schelter, Skew group rings, Houston J. Math., (to appear).

V. K. Kharchenko, Generalized identities with automorphisms, Algebra i Logika, 14 (1975), 215-237, (Engl. trans. March, 1976).

W. S. Martindale, Prime rings satisfying a generalized polynomial identity, J. Algebra 12 (1969), 576-584.

S. A. Amitsur, On rings of quotients, Symposia Mathematica 8 (1972), 149-164.

D. S. Passman, The algebraic structure of group rings, Wiley-Interscience, New York, 1977.

J. Lambek, Lectures on rings and modules, Blaisdell, Waltham, Mass., 1966.

A. Reid, Semiprime twisted group rings, J. London Math. Soc., (2), 12 (1976), 413-418.

BISERIAL RINGS [1]

Kent R. Fuller

The University of Iowa
Iowa City, Iowa 52242

A module is called uniserial (or often simply serial) in case its lattice of submodules is a chain. An artinian ring is called serial (or generalized uniserial) in case each of its indecomposable projective left and right modules is uniserial. Serial rings, introduced by Nakayama [18], were among the first studied non-semisimple rings with only finitely many indecomposable modules. An artinian ring with this latter property is said to be of finite (module or representation) type. Nakayama [18] proved that if R is a serial ring with radical J and basic set of primitive idempotents $e_1, \cdots, e_n$ (so that $Re_1, \cdots, Re_n$ are the left indecomposable projective R-modules) then every indecomposable left R-module is isomorphic to one of the form $Re_i/J^k e_i$; and similarly on the right.

An indecomposable module M is called biserial in case M contains uniserial submodules K_1 and K_2 such that $K_1 + K_2$ is M or the largest proper submodule of M and $K_1 \cap K_2$ is $\{0\}$ or the smallest nonzero submodule of M. This concept is self-dual because in the definition one may replace "K_i uniserial" by "M/K_i uniserial." A module with a largest proper (smallest nonzero) submodule is called local (colocal). Thus biserial modules are either local or colocal. We say that a ring is biserial in case each of its indecomposable projective left and right modules is biserial. Serial rings are biserial, and so are several other classes of rings of finite type. These include Tachikawa's algebras over which every indecomposable module is local or colocal, group algebras of finite type and rings over which every module is a direct sum of modules with distributive lattices of submodules. This article is principally a survey of the literature on these rings.

1. Preliminaries.

Let R be an artinian ring. The Auslander-Bridger transpose provides a correspondence T between finitely generated left and right R-modules with no projective direct summands such that

$$TT(M) \cong M$$
$$T(M_1 \oplus M_2) \cong T(M_1) \oplus T(M_2).$$

In particular T provides a 1-1 correspondence between

the finitely generated nonprojective indecomposable left and right R-modules. If M is a finitely generated R-module with minimal projective presentation

$$P_1 \xrightarrow{\gamma} P_0 \longrightarrow M \longrightarrow 0$$

then the transpose of M has minimal projective presentation

$$P_0^* \xrightarrow{\gamma^*} P_1^* \longrightarrow T(M) \longrightarrow 0,$$

where $(\ )^*$ is the functor $\mathrm{Hom}_R(\_,R)$. The properties of T are derived from this definition. (See [26] for a concise account.)

For a simple method of calculating T(M), write direct sums of left (right) modules as row (column) vectors and let homomorphisms operate on the opposite side of modules. Let $f_1,\cdots,f_m$ and $e_1,\cdots,e_m$ be primitive idempotents in R such that M has a minimal presentation

$$Rf_1 \oplus \cdots \oplus Rf_m \xrightarrow{A} Re_1 \oplus \cdots \oplus Re_n \longrightarrow M \longrightarrow 0$$

where $A = [\![a_{ij}]\!]$ with $a_{ij} \in f_i Re_j$. Then

1.1. $$T(M) \cong \begin{bmatrix} f_1R \\ \oplus \\ \vdots \\ \oplus \\ f_mR \end{bmatrix} \Bigg/ A \begin{bmatrix} e_1R \\ \oplus \\ \vdots \\ \oplus \\ e_nR \end{bmatrix}$$

The lattice of submodules of a module M is denoted by $\mathcal{L}(M)$. If $\mathcal{L}(M)$ is distributive we call M a <u>distributive module</u>. A modular lattice that is not distributive must have a sublattice of the form

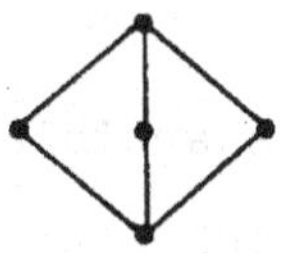

(see [4]). Using this fact (or see [5]) one can easily prove

1.2. PROPOSITION. *A module M is distributive if and only if each of its factor modules has a square free socle.*

(A semisimple module is square free iff it is zero or a direct sum of isomorphically distinct simple modules, i.e., iff it is distributive.)

Of course uniserial modules are distributive. The next proposition might be interpreted as saying that a module M of finite length is distributive in case $\mathcal{L}(M)$ is "naturally finite."

1.3. PROPOSITION. *Let ${}_RM$ be a module of finite length. If M is distributive then $\mathcal{L}(M)$ is finite. The converse is also true provided that the composition factors of M all have infinite endomorphism rings* (e.g., *if R is an algebra over an infinite field*).

Proof. The first statement is proved by a simple inducton argument. The second follows from (1.2) and the fact that if ${}_RS$ is simple then distinct elements $d \in End({}_RS)$ provide distinct submodules $\{(s,sd) \mid s \in S\}$ of $S \times S$.

In the remainder of this article, R is an associative (usually artinian) ring with identity. We denote the radical of R by $J = J(R)$, the injective envelope of a module M by E(M) and the socle of M by Soc M. If R is artinian then the upper Loewy series of ${}_RM$ is $M > JM > J^2M > \cdots > J^{\ell}M = 0$ and the lower Loewy series is $0 < Soc\ M < Soc^2\ M < \cdots < Soc^{\ell}\ M = M$ where $Soc^k\ M$ is the right annihilator $Soc^k\ M = r_M(J^k)$. If M has a composition series we write c(M) for its length.

2. Tachikawa Algebras.

Nakayama [18] and [19] proved that a ring is serial iff each of its indecomposable left and right modules is local. The first important result on biserial rings that we know of is due to Tachikawa. (Here an algebra is a finite dimensional algebra over a field, and we denote the dual module of M by D(M).)

2.1. THEOREM. [Tachikawa, 25] *An algebra* R *has the property that each of its indecomposable modules is either local or colocal if and only if for all primitive idempotents* e *and* f *in* R

(i) Je *and* eJ *are sums of* ≤ 2 *uniserial models; and*

(ii) *If* Je *is not uniserial and* Rf/Jf *embeds in a proper factor of* Je *then* fR *is uniserial; and the same holds for each* eJ *that is not uniserial.*

Over an artinian ring local (colocal) modules are those with unique maximal (minimal) submodules. Tachikawa said that the algebras of Theorem 2.1 were of cyclic-cocyclic module type. We shall simply call them *Tachikawa algebras*. These algebras are biserial as Tachikawa proved in [25, Proposition 2.7]. He also observed that (2.1) has the following

COROLLARY. ***Every indecomposable module over an algebra* R *is local if and only if* R *is left biserial and right serial.***

Tachikawa algebras have also been studied in [3] and [14], where they were characterized as those algebras over which every nonsimple indecomposable module has a nonzero proper submodule that is comparable to every other submodule. Here we shall discuss the structure of these algebras and their modules. In particular we shall show that their indecomposable modules can be obtained by repeated applications of the Auslander-Reiten "dual-of-the-transpose" function to a patently finite set of indecomposable modules.

Tachikawa used the various parts of the following lemma to prove the two implications of Theorem 2.1.

2.2. LEMMA. [Tachikawa, 25] *Let* R *be a Tachikawa algebra. Then*

(1) *If* M/J^2M *or* $\mathrm{Soc}^2\, M$ *is uniserial then so is* M;

(2) *If* M *and* N *are uniserial modules such that* $M/J^2M \cong N/J^2N$ $(\mathrm{Soc}^2\, M \cong \mathrm{Soc}^2\, N)$ *then one is an epimorph of (embeds in) the other*;

(3) *If* M *is both local and colocal then* M *is either uniserial or both projective and injective*.

Proof. See [25, Propositions (2.5), (4.1) and (3.1)] and their proofs.

Although the proof of the following theorem is lifted from Tachikawa's [25], it is not clear that he was aware that he actually had such a nice list of the indecomposable modules Observe that it follows from (2.3) that an algebra is a Tachikawa algebra if and only if all of its indecomposable modules are biserial.

2.3. THEOREM. [Tachikawa, 25] *Let R be a Tachikawa algebra with basic set of primitive idempotents* $e_1, \cdots, e_n$. *Then for* $i = 1, \cdots, n$

$$Je_i = K_{i1} + K_{i2} \quad \text{and} \quad e_iJ = L_{i1} + L_{i2}$$

with the K_{ij} *and* L_{ij} *uniserial or zero submodules such that* $K_{i1} \cap K_{i2} = \text{Soc Re}$ *or* 0 *and* $L_{i1} \cap L_{i2} = \text{Soc eR}$ *or* 0; *and every indecomposable R-module is isomorphic to or dual to one of the modules*

$$Re_i/(J^sK_{i1}+J^tK_{i2}) \quad \text{or} \quad e_iR/(L_{i1}J^s+L_{i2}J^t),$$

$i = 1, \cdots, n,\ \ s,t \geq 0$.

Proof. The K_{ij} and the H_{ij} exist since R is biserial [25, Proposition 2.7]. Suppose that $_RM$ is local, so that M is an epimorph of Re for some primitive idempotent $e \in R$. If Re is uniserial then so is M. If M is isomorphic to Re there is nothing to prove. Thus (factoring out Soc $_RR$ if Soc Re is simple) we may assume that $Je = K_1 \oplus K_2$ with K_i uniserial. Let

$$0 \longrightarrow N \xrightarrow{\subseteq} Re \xrightarrow{\gamma} M \longrightarrow 0$$

be exact, and observe that

$$JM = \gamma(K_1) + \gamma(K_2).$$

Suppose, as we may, that

$$c(\gamma(K_1)) = s \geq c(\gamma(K_2)).$$

Let $E = E(\gamma(K_1))$ be the injective envelope. Then by Tachikawa's theorem (2.1) E is uniserial, so $\gamma(K_1) = r_E(J^s)$ is injective over R/J^s. Thus, since $J^s\gamma(K_2) = 0$, we see that

$$JM = \gamma(K_1) \oplus \bar{K}_2$$

where $\bar{K}_2$ is an epimorph of $\gamma(K_2)$. Let $t = c(\bar{K}_2)$. Then $s \geq t$ and there are left ideals H_1 and H_2 with $N \leq H_i \leq Je$ such that

$$Je/N = H_1/N \oplus H_2/N,$$

$$H_1/N \cong K_1/J^sK_1 \quad \text{and} \quad H_2/N \cong K_2/J^tK_2.$$

Moreover, there is an isomorphism $\varphi : Re/(J^sK_1 \oplus K_2) \longrightarrow Re/H_2$, since both of these uniserial modules have length $s+1$ and injective envelope E. Now using the projectivity of Re we can find an automorphism Φ of Re such that the diagram

$$\begin{array}{ccc} Re & \xrightarrow{\Phi} & Re \\ \text{nat.}\downarrow & & \downarrow\text{nat.} \\ Re/(J^sK_1 \oplus K_2) & \xrightarrow{\varphi} & Re/H_2 \end{array}$$

commutes. But then, since $H_1 + H_2 = Je$ and $s \geq t$,

$$\Phi(J^sK_1) = J^s\Phi(K_1) \subseteq J^s(H_1+H_2)$$

$$\subseteq J^sH_1 + J^tH_2 \subseteq N;$$

and, by commutativity of the diagram,

$$\Phi(J^tK_2) = J^t\Phi(K_2) \subseteq J^tH_2$$

$$\subseteq N.$$

Thus, since they have the same composition length,

$$Re/(J^sK_1 \oplus J^tK_2) \cong Re/N \cong M.$$

If ${}_RM$ is a colocal module apply the above argument to its dual module $D(M)$.

Now we turn to the transposes of modules over Tachikawa algebras. Although we have no application for them here, it is worth noting that with a little more work one can explicitly calculate the transposes of the modules in the following two lemmas. In fact, if M is local with presentation as in 2.4 then M is a proper epimorph of $DT(M)$, if M is colocal but not local then $DT(M)$ embeds properly in M, and if $M \cong Re/\mathrm{Soc}\, Re$ with Re injective then $DT(M) \cong Je$.

2.4. LEMMA. <u>Let R be a Tachikawa algebra. If ${}_RM$ has a minimal presentation</u>

$$Rf_1 \oplus Rf_2 \longrightarrow Re \longrightarrow M \longrightarrow 0$$

<u>with e, f_i primitive idempotents in R then f_1R and f_2R are uniserial, and $T(M)$ is colocal with</u> $c(T(M)) > c(M)$.

<u>Proof</u>. By (2.3) we must have $M \cong Re/(J^sK_1 \oplus J^tK_2)$ with $0 \le s < c(K_1)$ and $0 \le t < c(K_2)$. So we can regard the presentation as

$$Rf_1 \oplus Rf_2 \xrightarrow{\begin{bmatrix} f_1x_1e \\ f_2x_2e \end{bmatrix}} Re \longrightarrow M \longrightarrow 0$$

with $f_1x_1e \in J^sK_1 \backslash J^{s+1}K_1 \subseteq J^{s+1} \backslash J^{s+2}$ and $f_2x_2e \in J^{t+1} \backslash J^{t+2}$. But then by (1.1)

$$T(M) \cong (f_1R \oplus f_2R)/(f_1x_1e, f_2x_2e)R.$$

Both f_1R and f_2R are uniserial by (2.1), and so, since $f_1x_1e \in J^{s+1}$ and $f_2x_2e \in J^{t+1}$, we have

$$c(T(M)) \geq c((f_1R\oplus f_2R)/(f_1x_1eR\oplus f_2x_2eR))$$

$$\geq s+1+t+1 = c(M)+1.$$

Since $T(M)$ is indecomposable and not local, $T(M)$ is colocal.

2.5. LEMMA. *Let* R *be a Tachikawa algebra. If* ${}_RM$ *is colocal but not local then* M *has a minimal presentation*

$$Re \longrightarrow Rf_1 \oplus Rf_2 \longrightarrow M \longrightarrow 0$$

with Re *primitive and* Rf_1 *and* Rf_2 *uniserial, and* $T(M)$ *is local with* $c(T(M)) < c(M)$.

Proof. If ${}_RM$ is colocal but not local then by (2.3) $D(M) \cong eR/(L_1J^s+L_2J^t)$ with $s \neq 0$ and $t \neq 0$. Let $f_1R/f_1J \cong L_1J^{s-1}/L_1J^s$ and $f_2R/f_2J \cong L_2J^{t-1}/L_2J^t$. Then by (2.1) Rf_1 and Rf_2 are uniserial. Moreover, $D(M)$ has an injective envelope $D(Rf_1) \oplus D(Rf_2) = E(D(M))$. Thus by duality there is an exact sequence $0 \longrightarrow K \xrightarrow{\subseteq} Rf_1 \oplus Rf_2 \longrightarrow M \longrightarrow 0$ with $K \subseteq Jf_1 \oplus Jf_2$. Suppose that K/JK is not simple. Then, since it is neither local nor colocal, $(Rf_1\oplus Rf_2)/JK = A_1 \oplus A_2$ with A_1 and A_2 local. Considering the projections on the A_1

we see that each A_i is an epimorph of Rf_1 or Rf_2 and so A_1 and A_2 are uniserial. But then $\mathrm{Soc}(A_1 \oplus A_2)$ $= \mathrm{Soc}\ A_1 \oplus \mathrm{Soc}\ A_2 = K/JK$ and so $M \cong R/K \cong A_1/\mathrm{Soc}\ A_1 \oplus A_2/\mathrm{Soc}\ A_2$, contrary to hypothesis. Thus K is local and we have the desired presentation

$$Re \longrightarrow Rf_1 \oplus Rf_2 \longrightarrow M \longrightarrow 0.$$

But then $T(M)$ has a minimal presentation

$$f_1R \oplus f_2R \longrightarrow eR \longrightarrow T(M) \longrightarrow 0$$

so $T(M)$ is local, and by Lemma 2.4, $c(M) = c(T(T(M)))$ $> c(T(M))$.

If M is a nonprojective left module over an algebra then $DT(M)$ is a noninjective left module and $TD(DT(M)) \cong M$. Thus DT defines a bijective function from the isomorphism classes of nonprojective left modules to the isomorphism classes of noninjective left modules over an algebra R. W. Müller [17] proved that over an algebra R with $J^2 = 0$ of finite type, every indecomposable module is of the form $DT^m(I)$ for some $m \geq 0$ with I either a simple or an injective module. If $\mathcal{I}$ is a finite set of indecomposable modules from which every indecomposable R-module can be calculated in this manner, we shall say that $\mathcal{I}$ is a

DT-basis for R. (Observe that $\mathcal{J}$ must always contain the indecomposable injective left R-modules.) In [21], Platzeck and Auslander proved that the indecomposable injective modules form a DT-basis for any hereditary algebra of finite type, and in [20], Platzeck generalized Müller's result by proving that his DT-basis works for any algebra stably equivalent to an hereditary algebra of finite type. Of course every algebra of finite type has a DT-basis. (We have no information about the converse[2] but it seems worth investigating.) But the idea is to find one whose modules (or their duals) are as easily described as possible. For Tachikawa algebras we have

2.6. PROPOSITION. *Let R be a Tachikawa algebra. Then R has a DT-basis consisting of injective modules and uniserial modules. Moreover, every nonsimple uniserial R-module is a submodule of a uniserial injective module or an epimorph of a uniserial projective module.*

Proof. The last statement follows at once from (2.1) and (2.3).

Suppose ${}_RM$ is colocal but not local. Then by (2.5) $DT(M)$ is colocal and $c(D(T(M))) = c(T(M)) < c(M)$, so there must be an integer m such that $(DT)^m(M)$ is local and colocal. Thus by (2.2.3) for each colocal module M there is an $m \geq 0$ such that $(DT)^m(M)$ is projective (and injective) or uniserial.

Now suppose that ${}_RM$ is local. Then by (2.3) M is either projective or has a minimal presentation of the form $Rf_1 \oplus Rf_2 \longrightarrow Re \longrightarrow M \longrightarrow 0$ or $Rf \longrightarrow Re \longrightarrow M \longrightarrow 0$. So by (2.4) there is an integer $k \geq 0$ such that $(DT)^k(M)$ is either projective or has a minimal presentation

$$0 \longrightarrow Rf \longrightarrow Re \longrightarrow (DT)^k(M) \longrightarrow 0.$$

In the latter case $T(DT)^k(M)$ is local, $(DT)^{k+1}(M)$ is colocal and by the preceding paragraph we have $m \geq k+1$ such that $(DT)^m(M)$ is projective or uniserial.

Of course, the above holds for right modules also. Now suppose that ${}_RM$ is an indecomposable left module that is not injective. Then for some $m \geq 1$, $D(TD)^m(M) = (DT)^m(D(M))$ is uniserial or projective (but not injective), $(TD)^m(M)$ is uniserial or injective (but not projective), and $M \cong (DT)^m(TD)^m(M)$, so the proof is complete.

According to (2.6) a Tachikawa algebra has a DT-basis $\mathcal{J}$ whose members are injective, simple, submodules of uniserial injective modules or factor modules of uniserial projective modules. Over any algebra such a set $\mathcal{J}$ is finite. Surely other interesting classes of algebras have such DT bases.

Müller [17] suggested the nonfaithful indecomposable modules (plus injectives) as a candidate for a DT-basis for an arbitrary ring of finite type. It is easy to see from the proof of (2.6) that a Tachikawa algebra actually has a DT-basis consisting of injective modules, simple modules, proper factors of uniserial projective modules, and proper submodules of uniserial injective modules. Thus Tachikawa algebras R (as well as algebras of finite type stably equivalent to hereditary algebras [17], [20], [21]) have DT-bases consisting of injective modules and modules over R/J^{ℓ} with $J^{\ell} \neq 0$. This is also trivially the case for any QF (or even QF-3) algebra. We wonder if this is true **for other interesting classes of rings of finite type.**

3. Diserial Rings.

We call a distributive biserial module a diserial module. An artinian ring whose indecomposable left and right projective modules are diserial is called a diserial ring. (We called them biserial in [12].) If R is a diserial ring with basic set of primitive idempotents $e_1, \cdots, e_n$, then for each $i = 1, \cdots, n$ there are unique uniserial left ideals K_{i1} and $K_{i2} \leq Je_i$ such that $K_{i1} + K_{i2} = Je_i$ and $K_{i1} \cap K_{i2}$ is zero or simple. We shall refer to these in the sequel.

Observe that by distributivity K_{i1} and K_{i2} have no common composition factors except possibly $\mathrm{Soc}\, K_{i1} = \mathrm{Soc}\, K_{i2}$. It is also worth noting that by [10, Lemma 1] the indecomposable injective modules over a diserial ring are diserial.

Janusz [15] showed that a split group algebra (e.g., one over an algebraically closed field) of finite type is diserial and that its structure and its indecomposable modules are determined by graphs used by Brauer and Dade to describe the characters in its blocks. These graphs are called Brauer trees.

In [11] we showed that each indecomposable weakly symmetric (i.e., QF with $\mathrm{Soc}\, Re_i \cong Re/Je_i$) diserial ring has an associated graph, with edges corresponding to a basic set of primitive idempotents $e_1,\cdots,e_n$ and vertices corresponding to the sets of composition factors of the various K_{ij}. (In this graph, edge e_i meets edge e_j iff $e_iRe_j \neq 0$.) Each vertex v in the graph is assigned a positive integer m_v. The vertex v is <u>exceptional</u> in case $m_v > 1$; and the graph is a <u>Brauer tree</u> in case it is a tree (has no closed paths) with at most one exceptional vertex. Janusz [15] completely described the indecomposable modules for split symmetric algebras of this form; and in [22] Reiten showed how to calculate the almost split exact sequences $0 \longrightarrow DT(M) \longrightarrow B \longrightarrow M \longrightarrow 0$ for each of their indecomposable modules.

3.1. THEOREM. _Let_ R _be an indecomposable weakly symmetric diserial ring. If_ R _is of finite type then its graph is a Brauer tree. If_ R _is a split symmetric algebra then the converse is true._

Proof. The first implication, proved in [11], uses the Dlab-Ringel generalization [7] of Gabriel's theorems on rings of finite type. The second is due to Janusz [15].

Nakayama [18] proved that every indecomposable module over a serial ring is uniserial. Fuller [9] and Skornjakov [23] proved that if every finitely generated indecomposable left module over an artinian ring is uniserial then the ring is serial. An artinian ring is said to be of (left) _distributive type_ in case each of its finitely generated indecomposable (left) modules is distributive. (An arbitrary ring is artinian of left distributive type iff each of its left modules is a direct sum of distributive modules [12, Remark (7.2)].) By (1.2) distributive type implies bounded type, so by Auslander's [1] a ring of left distributive type is of finite type. We do not know whether left distributive type implies right, but several results suggest that it may be so. In particular we were surprised to find in [10] that

3.2. THEOREM. _Every ring of left distributive type is diserial._

Of course if a ring has a duality between its finitely generated left modules and right modules and it is of left distributive type then it is of right distributive type. So for algebras or QF rings this question is moot. Even so, Theorem 3.2 has an application to weakly symmetric rings in the following theorem. For the proof see [11].

3.3. THEOREM. _Let R be an indecomposable weakly symmetric ring. If R is of distributive type then either R is serial or R is diserial and the graph of R is a Brauer tree with no exceptional vertex. If R is a split symmetric algebra then the converse is true._

For one-sided serial rings we proved [10, Corollary 6] the following result whose proof depends on the work of Tachikawa [24].

3.4. THEROEM. _Let R be a right serial ring. Then the following are equivalent:_

(a) _R is of left distributive type;_

(b) _Every indecomposable left R-module is local and distributive;_

(c) _R is diserial;_

(d) _Every indecomposable right R-module is colocal and distributive;_

(e) _R is of right distributive type._

We should note here that even though their structure is similar, algebras of distributive type need not be Tachikawa algebras. Indeed, most split symmetric algebras and most hereditary rings of distributive type have indecomposable modules that are neither local nor colocal.

The remaining results in this section are from some joint work with E. L. Green. The first gives another case in which left distributive type implies right. Before proceeding to it, we recall that the left quiver $\mathcal{Q}({}_R R)$ of an artinian ring R with basic set of idempotents $e_1,\cdots,e_n$ is constructed by drawing n vertices labeled $e_1,\cdots,e_n$ with u_{ij} arrows from the vertex labeled e_i to the one labeled e_j in case the Re_j/Je_j-th homogeneous component of Je_i/J^2e_j has length u_{ij}. (See [7] or [14].) The right quiver $\mathcal{Q}(R_R)$ is defined similarly.

3.5. PROPOSITION. Let R be an indecomposable hereditary artinian ring. Then the following are equivalent:

(a) R is of left distributive type;

(b) R is diserial of finite type;

(c) The quivers of R are Dynkin diagrams of type A_n;

(d) R is of right distributive type.

Proof. (a) ⇒ (b). By (3.2).

(b) ⇒ (c). According to the proof of [6, Proposition 10.2] any hereditary ring of finite type is isomorphic to a tensor ring as described in [7]. Any ring R over which Je/J^2e and eJ/eJ^2 are square free for each primitive idempotent $e \in R$ satisfies the dualization conditions described in [7, page 5]. Thus if R satisfies (b) then by Dlab and Ringel's [7] its quivers must be Dynkin diagrams. But using [13, Lemma 1] it is easy to check that an hereditary ring whose quivers are given by any of the Dynkin diagrams except A_n is not diserial.

(c) ⇒ (a). Let R be hereditary with quivers of type A_n. Then one can easily apply [10, Lemma 7] to construct $\frac{1}{2}n(n+1)$ indecomposable distributive left R-modules. But according to [7] and [14, page 131] R has only $\frac{1}{2}n(n+1)$ indecomposable left modules.

(c) ⇔ (d). By symmetry.

An indecomposable artinian ring is serial iff its left and right quivers are oriented cycles or oriented diagrams of type A_n.

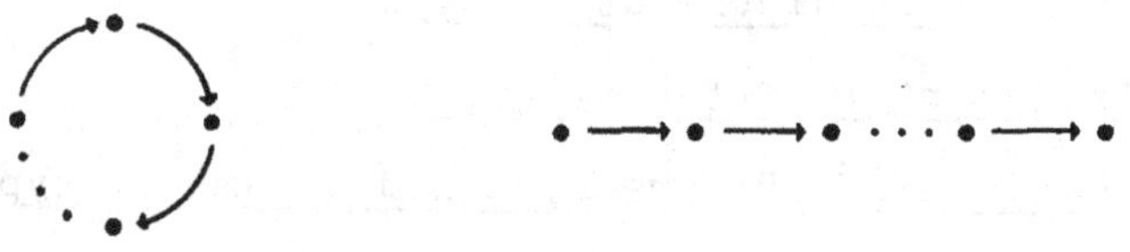

Any ring with (unoriented) quivers of type A_n is of distributive type because it is a factor ring of an hereditary ring with the same quivers [6, Prop. 10.2] (see also [13, Theorem 2]). As an application of (3.4) and (3.5), we shall next show that most rings whose quivers are cycles are also of distributive type. A source in a quiver is a vertex that arrows only leave and a sink is one that arrows only arrive at [7, page 1].

3.6. PROPOSITION. *Let R be an artinian ring with left and right quivers that are (unoriented) cycles. Let $e_1,\cdots,e_n$ be a basic set of idempotents numbered so that $e_1,\cdots,e_m$ correspond to the sources of $\mathcal{Q}({}_RR)$. Let $f = e_1 + \cdots + e_m$ and let $e = e_i$ correspond to a sink. If $eRf = 0$ then R is of distributive type.*

Proof. Given R, e and f as in the hypothesis observe that $Je = 0$ and $fJ = 0$. Let M be a left R-module and write (assuming that R is basic)

$$M = \Sigma_j Rg_j x_j + \Sigma_k Rh_k y_k$$

with $g_j, h_k \in \{e_1,\cdots,e_n\}$ such that

$$eRg_j x_j = 0 \quad \text{and} \quad eRh_j y_n \neq 0.$$

Let

$$K = \Sigma Rg_j x_j \quad \text{and} \quad L = \Sigma Rh_k y_k.$$

Since $eRf = 0$, no h_k belongs to $e_1,\cdots,e_m$. Thus

$fL \subseteq JL \cap fL = fJL = 0$, and so, since also $eK = 0$, we have

$$K = (1-e)K \quad \text{and} \quad L = (1-f)L.$$

In particular we may view L as a left $(1-f)R(1-f)$-module, and considering the left and right quivers of R we see that this ring is left serial and right diserial. Thus by (3.4) we may write

$$L = L' \oplus L''$$

where L' is a direct sum of colocal distributive modules whose socles are isomorphic to Re (recall $Je = 0$) and L'' a direct sum of ones whose socles are not isomorphic to Re. But then $eL'' = 0$ so we have

$$M = K' \oplus L'$$

with $K' = K + L'' = (1-e)K'$. But the quivers of $(1-e)R(1-e)$ are of type A_n, so by (3.5) K' is also a direct sum of distributive modules. Thus R is of left distributive type. By the proof of [6, Proposition 10.2] we see that R is a factor ring of an hereditary tensor ring T with the same quivers as R. From its quivers we see that the indecomposable projective modules over T are distributive. Thus T satisfies the dualization conditions [7, page 5], so by

[27, Proposition 5.8] T has a weakly symmetric self duality in the terminology of Haack's [28]. But then, also since the indecomposable projective modules over T are distributive, R has a self duality by [28, Proposition 4.1]. Thus R is of right, as well as left, distributive type.

Considering the quivers, it follows easily that every basic ring whose quivers are cycles has a factor by an ideal of left and right length ≤ 2 that is of distributive type.

[1] The author's research was partially supported by NSF Grant MCS77-00431.

[2] Idun Reiten has informed me that the converse is true, and follows from results in Auslander and Reiten's "Representation Theory of Artin Algebras, IV", Comm. in Algebra 5(1977), 443-518.

REFERENCES

[1] M. Auslander, Representation theory of artin algebras II, Comm. in Algebra 1(1974), 269-310.

[2] M. Auslander and M. Bridger, Stable module theory, Mem. Amer. Math. Soc. 94(1969).

[3] M. Auslander, E.L. Green and I. Reiten, Modules with waists, Illinois J. Math. 19(1975), 467-478.

[4] G. Birkhoff, Lattice theory, Amer. Math. Soc. Colloq. Publ. Vol. 25, 3rd Edition. Providence (1966).

[5] V.P. Camillo, Distributive modules, J. Algebra 36(1975), 16-25.

[6] V. Dlab and C.M. Ringel, On algebras of finite representation type, J. Algebra 33(1975), 306-394.

[7] V. Dlab and C.M. Ringel, Indecomposable representations of graphs and algebras, Mem. Amer. Math. Soc. No. 173, 6(1976).

[8] S.E. Dickson and K.R. Fuller, Algebras for which every indecomposable right module is invariant in its injective envelope, Pacific J. Math. 31(1969), 655-658.

[9] K.R. Fuller, On indecomposable injectives over artinian rings, Pacific J. Math. 29(1969), 115-135.

[10] K.R. Fuller, On a generalization of serial rings, Proc. of the Philadelphia Conference on Re. Thy., Dekker: Lect. Notes in Pure and Appl. Math. Vol. 37(1978), 359-368.

[11] K.R. Fuller, Weakly symmetric rings of distributive module type, Comm. in Algebra 5(1977), 997-1008.

[12] K.R. Fuller, Rings of left invariant module type, Comm. in Algebra 6(1978), 153-167.

[13] K.R. Fuller and J. Haack, Rings with quivers that are trees, Pacific J. Math., to appear.

[14] R. Gordon and E.L. Green, Modules with cores and amalgamations of indecomposable modules, Mem. Amer. Math. Soc. No. 187, 6(1976).

[15] G.J. Janusz, Indecomposable modules for finite groups, Ann. of Math. 89(1969), 209-241.

[16] G.J. Janusz, Some left serial algebras of finite type, J. Algebra 23(1972), 404-411.

[17] W. Müller, On artin rings of finite representation type, Proceedings of the International Conference on Representations of Algebras, Carleton University. Springer-Verlag: Lecture Notes in Math. 488(1975).

[18] T. Nakayama, On Frobeniusean algebras II, Ann. of Math. 42(1941), 1-22.

[19] T. Nakayama, Note on uniserial and generalized uniserial rings, Proc. Imp. Acad. Japan 16(1940), 285-289.

[20] M.I. Platzeck, Representation theory of algebras stably equivalent to an hereditary artin algebra, to appear.

[21] M.I. Platzeck and M. Auslander, Representation theory of hereditary artin algebras, Proc. of the Philadelphia Conference on Rep. Thy., Dekker: Lect. Notes in Pure and Appl. Math., Vol. 37(1978), 389-424.

[22] I. Reiten, Almost split sequences for group algebras of finite representation type, to appear.

[23] L.A. SKornjakov, When are all modules semi-chained?, Mat. Zametki 5(1969), 173-182.

[24] H. Tachikawa, On rings for which every indecomposable right module has a unique maximal submodule, Math. Z. 71(1959), 200-222.

[25] H. Tachikawa, On algebras of which every indecomposable representation has an irreducible one as the top or the bottom Loewy constituent, Math. Z. 75(1961), 215-227.

[26] R.B. Warfield, Jr., Serial rings and finitely presented modules, J. Algebra 37(1975), 187-222.

[27] M. Auslander, M.I. Platzeck and I. Reiten, Coxeter functors without diagrams, to appear.

[28] J. Haack, Self-duality and serial rings, to appear.

THE STATE SPACE OF K_0 OF A RING

K. R. Goodearl

The purpose of this paper is to introduce an invariant associated with any ring R, namely a compact convex set (known as the "state space") which is dual to the Grothendieck group $K_0(R)$, and to calculate this invariant in several cases. Our basic viewpoint is that the state space offers a compromise relative to K_0: the state space seems to be easier to compute in general than K_0 — in fact, the state space can be completely described in some situations in which calculating K_0 is hopeless — and while the state space does not carry as much information as K_0, in a number of situations the state space seems to carry enough information to be potentially useful. In that regard this study of state spaces is experimental, the object of the experiment being to investigate various rings R to determine what properties of R are perceived by the state space of $K_0(R)$, and in what form this information is stored in the state space. The present paper is intended as an introduction of state spaces to the mathematical community, as a report on the initial experimental investigation of state spaces, and as an invitation to participate in the experiment.

State spaces of Grothendieck groups have up to now been studied almost exclusively in the context of von Neumann regular rings, where they have been used quite successfully. For any regular ring R, there is a natural affine homeomorphism of the state space of $K_0(R)$ with the space $\mathbb{P}(R)$ of all pseudo-rank functions on R. This correspondence makes it quite easy, for example, to prove existence and uniqueness theorems about pseudo-rank functions [4, Chapter 18]. Also, the state space carries extensive information about finiteness properties of direct sums of projective modules, particularly when R is unit-regular.

For regular rings R, the state space of $K_0(R)$ is a certain kind of infinite-dimensional simplex. However, for other rings the state space can be arbitrary: every compact convex set appears as the state space of K_0 of a ring (Theorem 5). For HNP (hereditary noetherian prime) rings R; we present a complete description of the state space of $K_0(R)$ in Theorem 8. This description exhibits the state space as a certain amalgamation of simplexes associated with idempotent maximal ideals of R. For instance, the two- and three-dimensional possibilities for this state space are triangles, plane quadrilaterals, tetrahedrons, hexahedrons, octahedrons, and quadrilateral-base pyramids.

Most of the results of this paper are the product of joint work with R. B. Warfield, Jr., particularly the calculations over HNP rings, which are based on the results of [5]. Since the details of the HNP case are quite involved and are intended for inclusion in a separate joint paper, these details have been omitted here. This research was partially supported by grants from the National Science Foundation.

All rings in this paper are associative with unit, and all modules are unital right modules. We use $M_n(R)$ to denote the ring of all $n \times n$ matrices over a ring R. Given a module A, we use nA to denote the direct sum of n copies of A, and by the <u>rank</u> of A we mean the Goldie (uniform) dimension of A. If A has a composition series, we use $\ell(A)$ to denote the length of A.

I. BASIC CONCEPTS.

<u>DEFINITION</u>. We recall the description of the Grothendieck group $K_0(R)$ of a ring R. This group is an (additive) abelian group with a

generator $[A]$ for each finitely generated projective right R-module A, and with relations $[A \oplus B] = [A] + [B]$ for all such A and B. Generators $[A]$ and $[B]$ are equal in $K_0(R)$ if and only if $A \oplus nR \cong B \oplus nR$ for some positive integer n. Every element of $K_0(R)$ has the form $[A] - [B]$ for suitable A and B.

In addition to the abelian group structure on $K_0(R)$, there is a natural pre-order (i.e., a reflexive, transitive relation), which is compatible with the group structure. In order to discuss the combined structure, we require the following terminology.

DEFINITION. A cone in an abelian group G is an additively closed subset C such that $0 \in C$. We say that C is a strict cone if 0 is the only element $x \in G$ such that both $x \in C$ and $-x \in C$. Any cone C in G determines a pre-order $\leq_C$ on G, where $x \leq_C y$ if and only if $y - x \in C$. Note that $\leq_C$ is a partial order if and only if C is strict. In general, $\leq_C$ is always translation-invariant, that is, $x \leq_C y$ implies $x + z \leq_C y + z$. Conversely, any translation-invariant pre-order $\leq$ on G arises in this manner, from the cone $\{x \in G \mid x \geq 0\}$.

DEFINITION. A pre-ordered abelian group is a pair $(G, \leq)$, where G is an abelian group and $\leq$ is a particular translation-invariant pre-order on G. When there is no danger of confusion as to which pre-order is being used on G, we refer to G itself as a pre-ordered abelian group, and we write G^+ for the cone $\{x \in G \mid x \geq 0\}$. A partially ordered abelian group is a pre-ordered abelian group whose pre-order is a partial order.

DEFINITION. An order-unit in a pre-ordered abelian group G is an element $u \in G^+$ such that for any $x \in G$, there is a positive integer n for which $x \leq nu$. For example, using the normal partial order on $\mathbb{Z}$, every positive integer is an order-unit in $\mathbb{Z}$. Note that if there is an order-unit

$u \in G$, then any element $x \in G$ can be written as the difference of two elements of G^+, namely $x = nu - (nu - x)$, where n is a positive integer such that $x \leq nu$.

<u>DEFINITION</u>. For any ring R, we define $K_0(R)^+$ to be the collection of all generators $[A]$, where A is any finitely generated projective right R-module, and we note that $K_0(R)^+$ is a cone in $K_0(R)$. Then we equip $K_0(R)$ with the translation-invariant pre-order determined by this cone. Explicitly, we have $[A] - [B] \leq [C] - [D]$ in $K_0(R)$ if and only if $A \oplus D \oplus E \oplus nR \cong B \oplus C \oplus nR$ for some E and some positive integer n. Thus $K_0(R)$ becomes a pre-ordered abelian group. There is a natural order-unit in $K_0(R)$, namely $[R]$. Of course, if A is any finitely generated projective generator, then $[A]$ is an order-unit in $K_0(R)$ as well.

While the pre-order on $K_0(R)$ is not always a partial order, [4, Example 15.4], this does hold under mild finiteness restrictions on R. For example, $K_0(R)$ is partially ordered if R has finite rank on either side, or if R is commutative. More generally, the following condition may be used.

<u>DEFINITION</u>. A ring R is <u>directly finite</u> provided all one-sided inverses in R are two-sided, that is, $xy = 1$ implies $yx = 1$. As is easily checked, the endomorphism ring of a module A is directly finite if and only if A is not isomorphic to any proper direct summand of itself.

<u>PROPOSITION 1</u>. If R is a ring such that $M_n(R)$ is directly finite for all n, then the pre-order on $K_0(R)$ is a partial order.

<u>Proof</u>. It suffices to show that the cone $K_0(R)^+$ is strict. Thus consider any $[A] \in K_0(R)^+$ for which $-[A] \in K_0(R)^+$, so that $-[A] = [B]$ for

some B. Then $[A \oplus B] = 0$, hence $A \oplus B \oplus nR \cong nR$ for some positive integer n. Inasmuch as $M_n(R)$ is directly finite, nR is not isomorphic to any proper direct summand of itself, from which we see that $A \oplus B = 0$. Thus $A = 0$, so that $[A] = 0$. Therefore $K_0(R)^+$ is a strict cone. □

The concept of a pre-ordered abelian group with order-unit is an algebraic generalization of the standard functional-analytic concept of an <u>order-unit space</u>, i.e., a partially ordered real vector space with an order-unit (such a space is usually also assumed to be Archimedean, and is equipped with a norm derived from the partial order and the order-unit). For an order-unit space, the natural dual object to consider consists of linear functionals which respect the partial order and map the order-unit to the number 1. Such functionals are called <u>states</u>, the terminology being derived third-hand from quantum mechanics via the algebra of bounded operators on a Hilbert space and then via C*-algebras.

This terminology is carried one step further by extending the definition of states to pre-ordered abelian groups with order-unit, as follows.

<u>DEFINITION</u>. Let G be a pre-ordered abelian group with an order-unit u. A <u>state on</u> (G,u) is any order-preserving additive map $f : G \to \mathbb{R}$ such that $f(u) = 1$. (Note that an additive map $f : G \to \mathbb{R}$ is order-preserving if and only if $f(G^+) \subseteq \mathbb{R}^+$.) The <u>state space of</u> (G,u), denoted S(G,u), is the set of all states on (G,u). We view S(G,u) as a subset of the real vector space $\mathbb{R}^G$. Also, we equip $\mathbb{R}^G$ with the product topology, and we give S(G,u) the relative topology.

<u>PROPOSITION 2</u>. If G is a pre-ordered abelian group with an order-unit u, then S(G,u) is a compact convex subset of $\mathbb{R}^G$.

Proof. It is trivial to check that $S(G,u)$ is convex.

For each $x \in G$, there is a positive integer n_x such that $-n_x u \leq x \leq n_x u$. Note that the set

$$W = \{f \in \mathbb{R}^G \mid f(x) \in [-n_x, n_x] \text{ for all } x \in G\}$$

is compact, by Tychonoff's Theorem. Observing that $S(G,u)$ is a closed subset of W, we conclude that $S(G,u)$ is compact. □

Proposition 2 shows that the state space of any pre-ordered abelian group with order-unit is an object in an appropriate category of compact convex sets. Thus to completely describe such a state space in a given situation, we need to be able to specify it up to isomorphism in the appropriate category, as follows.

DEFINITION. Let K_1 and K_2 be compact convex subsets of linear topological spaces E_1 and E_2. Recall that a map $f: K_1 \to K_2$ is affine provided f preserves convex combinations. Equivalently, f is affine if and only if $f(\alpha x + (1-\alpha)y) = \alpha f(x) + (1-\alpha)f(y)$ for all $x,y \in K_1$ and all $\alpha \in [0,1]$. An affine homeomorphism is an affine map which is also a homeomorphism. If there exists an affine homeomorphism of K_1 onto K_2, then we say that K_1 and K_2 are affinely homeomorphic.

Thus we finally have sufficient terminology and concepts to describe the object of this paper:

THE INVARIANT: To any ring R, we associate the compact convex set $S(K_0(R),[R])$. (This is actually a contravariant functor from the category of rings to the category of compact convex sets, but we shall not need this.)

THE PROBLEM: Given a ring R, describe $S(K_0(R),[R])$ up to affine homeomorphism.

Some easy cases in which $S(K_0(R),[R])$ can be computed directly are given in Section II. In Section III we present a complete description of $S(K_0(R),[R])$ for any HNP ring R. This description is developed in terms of "extreme points", which play the role of vertices for arbitrary compact convex sets.

DEFINITION. Let K be a convex subset of a real vector space. An extreme point of K is any point of K which cannot be expressed as a nontrivial convex combination of two distinct points of K. In other words, a point $x \in K$ is extreme provided the only convex combinations $x = \alpha y + (1-\alpha)z$ with $y,z \in K$ and $\alpha \in [0,1]$ are those for which $\alpha = 0$, $\alpha = 1$, or $y = z$. For example, the extreme points of a convex polygon in the plane are just its vertices. While convex sets in general need not have any extreme points, compact convex sets are generated in a certain sense by their extreme points, as the following famous theorem shows.

KREIN-MILMAN THEOREM. [6, p. 131]. If K is a compact convex subset of a locally convex, Hausdorff, linear topological space, then K equals the closure of the convex hull of its extreme points. □

The Krein-Milman Theorem is even more explicit in case K has only finitely many extreme points $e_1,\ldots,e_n$. In this case, the convex hull of the e_i is compact (because it is a continuous image of the standard (n-1)-dimensional simplex, which is compact), hence K is just the convex hull of its extreme points. To complete the description of K then requires describing what affine relations (if any) exist among the e_i.

II. EXAMPLES.

Perhaps the easiest example to compute is the case of a semisimple ring R. Then $R \cong R_1 \times \dots \times R_k$ for some simple artinian rings R_i of length n_i. Choosing a simple right R_i-module A_i for each i, we see that $K_0(R)$ is a free abelian group with basis $\{[A_1],\dots,[A_k]\}$, that $K_0(R)^+ = \mathbb{Z}^+[A_1] + \dots + \mathbb{Z}^+[A_k]$, and that $[R] = n_1[A_1] + \dots + n_k[A_k]$. Thus $K_0(R)$ is isomorphic to $\mathbb{Z}^k$, equipped with the componentwise partial order (so that $(x_1,\dots,x_k) \leq (y_1,\dots,y_k)$ if and only if $x_i \leq y_i$ for all i), and with the order-unit $u = (n_1,\dots,n_k)$. Then $S(K_0(R),[R])$ is affinely homeomorphic to $S(\mathbb{Z}^k,u)$. There is a natural isomorphism of $\mathbb{R}^k$ onto $\mathrm{Hom}_{\mathbb{Z}}(\mathbb{Z}^k,\mathbb{R})$, where $(\alpha_1,\dots,\alpha_k)$ corresponds to the map $(x_1,\dots,x_k) \mapsto \alpha_1 x_1 + \dots + \alpha_k x_k$. Under this isomorphism, $S(\mathbb{Z}^k,u)$ corresponds to the set

$$\{(\alpha_1,\dots,\alpha_k) \in \mathbb{R}^k \mid \text{all } \alpha_i \geq 0 \text{ and } \alpha_1 n_1 + \dots + \alpha_k n_k = 1\},$$

which is affinely homeomorphic to the standard $(k-1)$-dimensional simplex. Thus we conclude that $S(K_0(R),[R])$ is affinely homeomorphic to the standard $(k-1)$-dimensional simplex.

Since any abelian group which has $\mathbb{Z}$ as a direct summand can be realized as K_0 of some Dedekind domain, it is plausible to expect a wide variety of compact convex sets as state spaces of K_0 of Dedekind domains. In fact, the opposite is true: the state space of K_0 of any Dedekind domain R is a single point, as we show in Theorem 4. Roughly speaking, the reason is that finitely generated projective R-modules are "approximately free" (since any such module is a direct sum of a free module and an ideal of R), which leaves no room for variation in the values of states on $K_0(R)$.

LEMMA 3. Let R be a commutative integral domain, and let A be a finitely generated projective R-module. Then there exist nonnegative integers $d, n_1, n_2, \ldots$ and R-modules $B_1, B_2, \ldots$ such that $kA \cong n_k R \oplus B_k$ and $\operatorname{rank}(B_k) \leq d$ for all $k = 1,2,\ldots$.

Proof. Choose a finitely generated subring S of R such that $A \cong C \otimes_S R$ for some finitely generated projective S-module C. Since S is a homomorphic image of a polynomial ring over $\mathbb{Z}$ in finitely many indeterminates, S has finite Krull dimension d.

Let k be a positive integer. If $\operatorname{rank}(kC) > d$, then for all maximal ideals M of S we see that kC_M is free of rank at least $d+1$, whence $(d+1)S_M$ is a direct summand of kC_M. In this case, Serre's Theorem [9, Theorem 11.2] says that $kC \cong S \oplus C_1$ for some C_1. If $\operatorname{rank}(C_1) > d$, then similarly $C_1 \cong S \oplus C_2$ for some C_2. Since this process must eventually stop, we obtain $kC \cong n_k S \oplus D_k$ for some nonnegative integer n_k and some S-module D_k such that $\operatorname{rank}(D_k) \leq d$.

Set $B_k = D_k \otimes_S R$, so that $kA \cong n_k R \oplus B_k$. If Q is the quotient field of R, then the quotient field P of S is a subfield of Q, hence $B_k \otimes_R Q \cong D_k \otimes_S Q \cong (D_k \otimes_S P) \otimes_P Q$. Thus

$$\operatorname{rank}(B_k) = \dim_Q(B_k \otimes_R Q) = \dim_P(D_k \otimes_S P) = \operatorname{rank}(D_k) \leq d. \quad \square$$

THEOREM 4. If R is any commutative integral domain, then $S(K_0(R),[R])$ consists of a single point.

Proof. Note that the rule $r([A]-[B]) = \operatorname{rank}(A) - \operatorname{rank}(B)$ defines a state $r \in S(K_0(R),[R])$. We must show that any $s \in S(K_0(R),[R])$ equals r.

If A is any finitely generated projective R-module, then by Lemma 3 there exist nonnegative integers $d, n_1, n_2, \ldots$ and R-modules $B_1, B_2, \ldots$ such that $kA \cong n_k R \oplus B_k$ and $\operatorname{rank}(B_k) \leq d$ for all $k = 1,2,\ldots$. Note that

$r([A]) \leq (n_k + d)/k$. We also have $k \cdot s([A]) = n_k + s([B_k]) \geq n_k$, whence $s([A]) \geq n_k/k \geq r([A]) - (d/k)$. Since this holds for all k, we obtain $s([A]) \geq r([A])$.

Now $A \oplus B \cong nR$ for some B and some positive integer n. Repeating the argument above, we obtain $s([B]) \geq r([B])$, whence $s([A]) = n - s([B]) \leq n - r([B]) = r([A])$, and consequently $s([A]) = r([A])$.

Therefore $s = r$. □

The state space of K_0 of a von Neumann regular ring R has been studied extensively, for in this case $S(K_0(R),[R])$ is affinely homeomorphic to the compact convex set $\mathbb{P}(R)$ of all pseudo-rank functions on R [4, Proposition 17.12]. By [4, Theorem 17.5], $\mathbb{P}(R)$ is a _Choquet simplex_, which is an infinite-dimensional generalization of classical simplexes. We will not discuss Choquet simplexes here except to say that every finite-dimensional Choquet simplex is affinely homeomorphic to a standard finite-dimensional simplex [8, Proposition 9.11]. Given any metrizable Choquet simplex K, there exists a simple regular ring R such that K is affinely homeomorphic to $\mathbb{P}(R)$ and thus to $S(K_0(R),[R])$ [4, Theorem 17.23].

In the examples discussed so far, the state spaces have been simplexes of one kind or another. In general, however, there is no restriction on the kind of compact convex sets which can be state spaces of K_0 of rings, as the following theorem shows.

THEOREM 5. Let K be a compact convex subset of a locally convex, Hausdorff, linear topological space. Then there exists a right and left semihereditary ring R such that $S(K_0(R),[R])$ is affinely homeomorphic to K.

Proof. Let $A(K)$ denote the Banach space of all affine continuous real-valued functions on K. There is a natural partial order on $A(K)$, where $f \leq g$ if and only if $f(x) \leq g(x)$ for all $x \in K$. Using this partial order, $A(K)$ becomes a partially ordered vector space, and the constant function 1 is an order-unit in $A(K)$. According to [1, Theorem II.2.4] or [11, Theorem 23.2.3], K is affinely homeomorphic to the state space of $(A(K),1)$. (These theorems refer to the functional-analytic state space of $(A(K),1)$; in order to see that this is the same as the state space as we have defined it, we must check that all states in $S(A(K),1)$ are $\mathbb{R}$-linear.) In view of [2, Theorems 6.2, 6.4], there exists a right and left semihereditary ring R such that $(K_0(R),[R]) \cong (A(K),1)$, that is, there is a group isomorphism of $K_0(R)$ onto $A(K)$ which is also an order-isomorphism and which maps $[R]$ to 1. Consequently, $S(K_0(R),[R])$ is affinely homeomorphic to $S(A(K),1)$ and thus to K. □

In particular, Theorem 5 says that there exist rings R for which $S(K_0(R),[R])$ is not a simplex. However, it gives no indication of whether this could happen with any of the rings one might run across in practice. To show that this does indeed happen, we now present an example in which it can be calculated directly that $S(K_0(R),[R])$ is a square. This example is a particular case of our general results for state spaces of K_0 of HNP rings in the next section, but the calculations here are much more direct, in that we obtain $S(K_0(R),[R])$ from a precise description of $K_0(R)$.

Set $R = \begin{pmatrix} \mathbb{Z} & 6\mathbb{Z} \\ \mathbb{Z} & \mathbb{Z} \end{pmatrix}$. Observing that R is the idealizer of the semimaximal right ideal $M = \begin{pmatrix} 6\mathbb{Z} & 6\mathbb{Z} \\ \mathbb{Z} & \mathbb{Z} \end{pmatrix}$ in $M_2(\mathbb{Z})$, we see by [10, Theorem 4.3] that R is an HNP ring. Now set

$$E = \begin{pmatrix} 2\mathbb{Z} & 6\mathbb{Z} \\ 0 & 0 \end{pmatrix} \quad ; \quad F = \begin{pmatrix} 3\mathbb{Z} & 6\mathbb{Z} \\ 0 & 0 \end{pmatrix} \quad ; \quad G = \begin{pmatrix} 0 & 0 \\ \mathbb{Z} & \mathbb{Z} \end{pmatrix} \quad ;$$

each of which is a right ideal of R and thus is a finitely generated projective right R-module. We note that $EM \cong FM \cong G$, that $GM = G$, and that $M \cong 2G$. In addition, $R/M \cong (E/EM) \oplus (F/FM)$, hence by Schanuel's Lemma we obtain

$$R \oplus 2G \cong R \oplus EM \oplus FM \cong E \oplus F \oplus M \cong E \oplus F \oplus 2G .$$

Thus $[R] = [E] + [F]$ in $K_0(R)$.

As far as the abelian group structure is concerned, we claim that $K_0(R)$ is free with basis $\{[E],[F],[G]\}$. Given any finitely generated projective right R-module A, we see that AM is a finitely generated projective right $M_2(\mathbb{Z})$-module, and consequently $AM \cong nG$ for some $n \geq 0$. Since R/M is a semisimple ring with exactly two isomorphism classes of simple right modules, represented by E/EM and F/FM, we must have $A/AM \cong s(E/EM) \oplus t(F/FM)$ for some $s,t \geq 0$. Applying Schanuel's Lemma again, we obtain

$$A \oplus (s+t)G \cong A \oplus s(EM) \oplus t(FM) \cong sE \oplus tF \oplus AM \cong sE \oplus tF \oplus nG ,$$

whence $[A] = s[E] + t[F] + (n-s-t)[G]$ in $K_0(R)$. Thus $[E],[F],[G]$ span $K_0(R)$.

Suppose that $m_1[E] + m_2[F] + m_3[G] = n_1[E] + n_2[F] + n_3[G]$ for some $m_i, n_i \geq 0$. Then

$$m_1 E \oplus m_2 F \oplus m_3 G \oplus kR \cong n_1 E \oplus n_2 F \oplus n_3 G \oplus kR$$

for some $k \geq 0$. Since $GM = G$, we obtain

$$m_1(E/EM) \oplus m_2(F/FM) \oplus k(R/M) \cong n_1(E/EM) \oplus n_2(F/FM) \oplus k(R/M) ,$$

from which we infer that $m_1 = n_1$ and $m_2 = n_2$. Then $m_3 G \oplus hR \cong n_3 G \oplus hR$ for some $h \geq 0$, whence $m_3 + 2h = \operatorname{rank}(m_3 G \oplus hR) = \operatorname{rank}(n_3 G \oplus hR) = n_3 + 2h$ and so $m_3 = n_3$. Thus $[E],[F],[G]$ are linearly independent over $\mathbb{Z}$.

Therefore $\{[E],[F],[G]\}$ is a basis for $K_0(R)$, as claimed. Observing that G is a direct summand of R, we see that $[G] \leq [R] = [E] + [F]$ in

$K_0(R)$, whence $[E]+[F]-[G] \in K_0(R)^+$. We claim that

$$K_0(R)^+ = \mathbb{Z}^+[E]+\mathbb{Z}^+[F]+\mathbb{Z}^+[G]+\mathbb{Z}^+([E]+[F]-[G]).$$

Given $[A] \in K_0(R)^+$, we infer as above that $A \oplus (s+t)G \cong sE \oplus tF \oplus nG$ for some $s,t,n \geq 0$. Set $K = \begin{pmatrix} \mathbb{Z} & 6\mathbb{Z} \\ \mathbb{Z} & 6\mathbb{Z} \end{pmatrix}$, which is a two-sided ideal of R, and note that R/K is a semisimple ring with exactly two isomorphism classes of simple right modules, represented by E/EK and F/FK. Note also that $G/GK \cong R/K \cong (E/EK) \oplus (F/FK)$. Then

$$(A/AK) \oplus (s+t)((E/EK) \oplus (F/FK)) \cong s(E/EK) \oplus t(F/FK) \oplus n((E/EK) \oplus (F/FK)) ,$$

from which it follows that $s \leq n$ and $t \leq n$. Now

$$\begin{aligned} [A] &= s[E]+t[F]+(n-s-t)[G] \\ &= (t-s)[F]+(n-t)[G]+s([E]+[F]-[G]) \qquad (1) \\ &= (s-t)[E]+(n-s)[G]+t([E]+[F]-[G]) . \qquad (2) \end{aligned}$$

If $s \leq t$, then the coefficients in (1) are nonnegative, while if $s \geq t$, then the coefficients in (2) are nonnegative. In either case, we have an expression for $[A]$ as a nonnegative linear combination of $[E],[F],[G]$, $[E]+[F]-[G]$. Thus $K_0(R)^+ = \mathbb{Z}^+[E]+\mathbb{Z}^+[F]+\mathbb{Z}^+[G]+\mathbb{Z}^+([E]+[F]-[G])$, as claimed.

Therefore we find that $K_0(R)$ is isomorphic to the abelian group $\mathbb{Z}^3$ with the partial order obtained from the cone

$$\mathbb{Z}^+(1,0,0)+\mathbb{Z}^+(0,1,0)+\mathbb{Z}^+(0,0,1)+\mathbb{Z}^+(1,1,-1)$$

and with the order-unit $(1,1,0)$. As a result, $S(K_0(R),[R])$ is affinely homeomorphic to the set

$$K = \{(\alpha_1,\alpha_2,\alpha_3) \in \mathbb{R}^3 \mid \text{all } \alpha_i \geq 0; \;\; \alpha_1+\alpha_2 \geq \alpha_3; \;\; \alpha_1+\alpha_2 = 1\}.$$

Note that K is a rectangular subset of the plane $\alpha_1+\alpha_2 = 1$, so that K is affinely homeomorphic to a square.

Therefore $S(K_0(R),[R])$ is affinely homeomorphic to a square.

III. HNP RINGS.

Throughout this section, we assume that R is an HNP ring with maximal quotient ring Q, and we set $S = S(K_0(R),[R])$. We present a complete description of S in terms of the structure of R; more precisely, in terms of the idempotent maximal ideals of R. One feature of this description is that S is infinite-dimensional if and only if R has infinitely many idempotent maximal ideals. It is a longstanding open question whether there exist HNP rings with infinitely many idempotent maximal ideals, but we conjecture that such rings do exist. For all known examples, R has only finitely many idempotent maximal ideals, and for these rings the description of S given below simplifies accordingly.

DEFINITION. Recall that if M is any ideal of R, the right and left orders of M (in Q) are the rings

$$O_r(M) = \{x \in Q \mid Mx \subseteq M\} \qquad ; \qquad O_\ell(M) = \{x \in Q \mid xM \subseteq M\} .$$

A cycle of idempotent maximal ideals of R is any finite ordered set $\{M_1,\ldots,M_n\}$ of distinct idempotent maximal ideals such that $O_r(M_1) = O_\ell(M_2)$, $O_r(M_2) = O_\ell(M_3)$, ... , $O_r(M_n) = O_\ell(M_1)$. Such a cycle always contains at least two idempotent maximal ideals, since it is not possible to have $O_r(M_1) = O_\ell(M_1)$ for an idempotent maximal ideal M_1.

The existence of cycles depends on the existence of invertible ideals, for an idempotent maximal ideal M of R belongs to a cycle if and only if M contains an invertible ideal [3, Propositions 2.4, 2.5]. If R is right bounded, then every nonzero ideal of R contains an invertible ideal [7, Theorem 3.3], so in this case every nonzero idempotent maximal ideal of R belongs to a cycle.

For an example without cycles, let T be a simple HNP ring which is not artinian, and let R be the idealizer of a maximal right ideal M of T. Then R is an HNP ring by [10, Theorem 4.3], and the only ideals of R are $0,M,R$. Thus M is the unique maximal ideal of R, and M is idempotent, but M does not belong to a cycle (because R has no other idempotent maximal ideals).

DEFINITION. The easiest state to define on $K_0(R)$ is obtained by normalizing rank, to obtain a state $r \in S$ given by the rule $r([A]-[B]) = (\mathrm{rank}(A) - \mathrm{rank}(B))/\mathrm{rank}(R)$. More generally, we can construct similar states by using normalized ranks computed over factor rings of R. The only such states for which we need explicit notation are those arising from nonzero idempotent maximal ideals of R. Namely, if M is any nonzero idempotent maximal ideal of R, then R/M is an artinian ring, and we define a state $e(M) \in S$ according to the rule

$$e(M)([A]-[B]) = (\ell(A/AM) - \ell(B/BM))/\ell(R/M) .$$

The basic relations among these states arise from cycles, as follows.

PROPOSITION 6. If $\{M_1,\ldots,M_n\}$ is a cycle of idempotent maximal ideals of R, then

$$r = (\ell(R/M_1)e(M_1) + \ldots + \ell(R/M_n)e(M_n))/(\ell(R/M_1) + \ldots + \ell(R/M_n)) .$$

Proof. This follows from the fact that

$$\mathrm{rank}(A)/\mathrm{rank}(R) = (\ell(A/AM_1) + \ldots + \ell(A/AM_n))/(\ell(R/M_1) + \ldots + \ell(R/M_n))$$

for all finitely generated projective right R-modules A, which was proved in [5, Corollary 34]. □

Thus if R has a cycle $\{M_1,\ldots,M_n\}$ of idempotent maximal ideals, then r is a nontrivial convex combination of the $e(M_i)$, so that r is not an extreme point of S. Otherwise, it turns out that r is an extreme point of S. The only other extreme points of S are those of the form $e(M)$, as follows.

THEOREM 7. Let E be the set of extreme points of S, and let

$$X = \{e(M) \mid M \text{ is a nonzero idempotent maximal ideal of } R\} .$$

(a) If R has cycles of idempotent maximal ideals, then $E = X$.

(b) If R has no cycles of idempotent maximal ideals, then $E = X \cup \{r\}$.

(c) X is a discrete subspace of S, and $X \cup \{r\}$ is a compact subspace of S.

Proof. We first prove that $X \subseteq E \subseteq X \cup \{r\}$.

Consider any $e(M) \in X$, and suppose that we have a convex combination $e(M) = \alpha s + (1-\alpha)t$ with $0 < \alpha < 1$ and $s,t \in S$. Since $e(M)([M]) = 0$, we see that $s([M]) = 0$. Likewise, if A is any finitely generated projective right R-module, then $s([AM]) = 0$. Note that $n(A/AM) \cong k(R/M)$, where $n = \ell(R/M)$ and $k = \ell(A/AM)$. Then $nA \oplus kM \cong kR \oplus n(AM)$ (by Schanuel's Lemma), whence $ns([A]) = k$, and consequently $s([A]) = k/n = e(M)([A])$. Thus $s = e(M)$, and likewise $t = e(M)$, which shows that $e(M)$ is an extreme point of S.

Therefore $X \subseteq E$.

Now consider any $s \in E$, and suppose that $s \neq r$. Then we must have $s([A]) < r([A])$ for some finitely generated projective right R-module A. Since A is isomorphic to a direct sum of uniform right ideals of R, we obtain $s([B]) < r([B])$ for some uniform right ideal B of R. There exists an essential right ideal C of R which is isomorphic to a direct sum of copies of B, whence $s([C]) < r([C]) = 1$.

Inasmuch as R/C has a composition series, we infer from this that $s([K]) < 1$ for some maximal right ideal K of R.

If R/K is faithful, then every finite direct sum of copies of R/K must be cyclic. Consequently, R has right ideals $K_1, K_2, \ldots$ such that each $R/K_n \cong n(R/K)$. By Schanuel's Lemma, $R \oplus nK \cong nR \oplus K_n$, whence $1 + ns([K]) = n + s([K_n]) \geq n$. But then $s([K]) \geq (n-1)/n$ for all n, which contradicts the fact that $s([K]) < 1$.

Thus R/K must be unfaithful, hence the annihilator of R/K is a nonzero maximal ideal M of R. Since $s([K]) < 1$, we compute that $s([M]) < 1$.

If M is invertible, then for each positive integer n, right multiplication by M^n induces an isomorphism between the submodule lattices of R/M and M^n/M^{n+1}. As a result, we infer that $R/M \cong M^n/M^{n+1}$ (as right R-modules), and then using Schanuel's Lemma again, we obtain $[R] - [M] = [M^n] - [M^{n+1}]$ in $K_0(R)$. Adding these equations for $n = 1, \ldots, k$, we obtain $k[R] - k[M] = [M] - [M^{k+1}]$, whence $k[R] + [M^{k+1}] = (k+1)[M]$. But then $(k+1)s([M]) = k + s([M^{k+1}]) \geq k$ for all k, which contradicts the fact that $s([M]) < 1$.

Thus M is not invertible, hence M must be idempotent. If $s([M]) = 0$, then we compute that $s([AM]) = 0$ for all finitely generated projective right R-modules A. In this case, it follows as above that $s = e(M) \in X$. Now assume that $s([M]) \neq 0$.

Set $\alpha = s([M])$, so that $0 < \alpha < 1$, and define $t \in S$ by the rule $t([A]) = s([AM])/\alpha$. We claim that $(s - \alpha t)([A]) \geq 0$ for all finitely generated projective right R-modules A. Note that $n(A/AM) \cong k(R/M)$, where $n = \ell(R/M)$ and $k = \ell(A/AM)$. Then $nA \oplus kM \cong kR \oplus n(AM)$, hence $ns([A]) + k = k + n\alpha t([A])$, and consequently $n(s-\alpha t)([A]) = k(1-\alpha) \geq 0$. Thus $(s-\alpha t)([A]) \geq 0$, as claimed. As a result, we find that $s-\alpha t = (1-\alpha)t'$ for some $t' \in S$. Observing that $t'([M]) = 0$, we infer as above that $t' = e(M)$, hence $s = \alpha t + (1-\alpha)e(M)$. Inasmuch as s is an extreme point of S, we conclude that $s = e(M) \in X$.

Therefore $E \subseteq X \cup \{r\}$.

(a) It is clear from Proposition 6 that $r \notin E$ in this case, hence $E = X$.

(b) If $r \notin E$, then there is a convex combination $r = \alpha s + (1-\alpha)t$ with $0 < \alpha < 1$ and distinct $s, t \in S$. Note that $s \neq r$. Proceeding as above, we obtain $s([M]) < 1$ for some nonzero idempotent maximal ideal M of R, and then either $s = e(M)$ or s is a positive convex combination of $e(M)$ and some other state. In either case, there is a convex combination $r = \beta e(M) + (1-\beta)t'$ with $0 < \beta < 1$ and $t' \in S$. Consequently, we compute that $\ell(A/AM) \leq \ell(R/M)/\beta$ for all right ideals A of R. But then [5, Theorem 35] says that M belongs to a cycle, which is false.

Therefore $r \in E$ in this case, whence $E = X \cup \{r\}$.

(c) First consider any net $\{s_i\}$ in $X \cup \{r\}$ which converges to some $s \in S$, and assume that $s \neq r$. In particular, $\{i \mid s_i = r\}$ is not cofinal in the index set, hence there is an index j such that $s_i = e(M_i) \in X$ for all $i > j$. Since $s \neq r$, we infer as above that $s([M]) < 1$ for some nonzero idempotent maximal ideal M of R. Then there is an index $k > j$ such that $s_i([M]) < 1$ for all $i > k$. Consequently, we infer that $M_i = M$ for all $i > k$, whence $s = e(M) \in X$.

Therefore $X \cup \{r\}$ is closed in S and so is compact.

If X is not discrete, then there is a net $\{s_i\}$ in X which converges to some $s \in X$ such that all $s_i \neq s$. We have $s = e(M)$ for some nonzero idempotent maximal ideal M of R, hence $s([M]) = 0 < 1$. But then we infer as above that there exists an index k such that $s_i = e(M)$ for all $i > k$, which is false.

Therefore X is discrete. □

Because of the Krein-Milman Theorem, it follows from Theorem 7 that S is generated (as a compact convex set) by r and the $e(M)$. It can be shown that the only finite affine relations among r and the $e(M)$ are consequences of the relations given by Proposition 6. Thus S may be described as the closure (in a suitable space) of a convex set constructed by amalgamating simplexes corresponding to cycles of idempotent maximal ideals of R and adding affinely independent points corresponding to nonzero idempotent maximal ideals of R which do not belong to cycles. A precise description of S is given in the following theorem.

THEOREM 8. Let I_1 and I_2 be disjoint index sets, let $\{X_i \mid i \in I_1\}$ be the collection of cycles of idempotent maximal ideals of R, let $\{M_i \mid i \in I_2\}$ be the collection of nonzero idempotent maximal ideals of R which do not belong to cycles, and set $I = I_1 \cup I_2$.

For $i \in I_2$, set $E_i = \mathbb{R}$ and $K_i = [0,1]$.

For $i \in I_1$, list $X_i = \{M_{i1}, \ldots, M_{i,n(i)}\}$ and set $E_i = \mathbb{R}^{n(i)-1}$. For $j = 1, \ldots, n(i)-1$, define $e_{ij} \in E_i$ so that the j-th. coordinate of e_{ij} is $1/\ell(R/M_{ij})$ and all other coordinates of e_{ij} are zero. Define $e_{i,n(i)} \in E_i$ so that all coordinates of $e_{i,n(i)}$ are $-1/\ell(R/M_{i,n(i)})$. Let K_i be the convex hull of $\{e_{i1}, \ldots, e_{i,n(i)}\}$.

Set $E = \prod_{i \in I} E_i$, and give E the product topology. For all $i \in I$, let K_i' be the set of those points of E for which the i-th. component lies in K_i and all other components are zero. Set K equal to the closure of the convex hull of $\bigcup_{i \in I} K_i'$.

Then S is affinely homeomorphic to K.

Sketch of Proof. The major difficulty of the proof is the construction of suitable projections from S onto each K_i.

Given $i \in I_1$, set $\lambda_{ij} = \ell(R/M_{ij})$ for $j = 1,\dots,n(i)$, and set $\lambda_i = \lambda_{i1} + \dots + \lambda_{i,n(i)}$. Also set $I_{ij} = (\bigcap_{k \neq j} M_{ik})^{n(i)-1}$ for $j = 1,\dots,n(i)-1$. Then we define an affine continuous map $\psi_i : S \to E_i$ by the rule $\psi_i(s) = (s([I_{i1}]),\dots,s([I_{i,n(i)-1}]))$. It can be computed that $\psi_i(e(M)) = \psi_i(r) = (1,1,\dots,1)$ for all nonzero idempotent maximal ideals $M \notin X_i$, that $\psi_i(e(M_{ij})) = (0,\dots,0,\lambda_i/\lambda_{ij},0,\dots,0)$ (with λ_i/λ_{ij} in the j-th. position) for all $j = 1,\dots,n(i)-1$, and that $\psi_i(e(M_{i,n(i)})) = 0$.

We check that $\psi_i(r),\psi_i(e(M_{i1})),\dots,\psi_i(e(M_{i,n(i)-1}))$ are affinely independent. Consequently, there exists an affine continuous map $\theta_i : E_i \to E_i$ such that $\theta_i\psi_i(r) = 0$ and $\theta_i\psi_i(e(M_{ij})) = e_{ij}$ for $j = 1,\dots,n(i)-1$. Also, we compute that $\theta_i\psi_i(e(M_{i,n(i)})) = e_{i,n(i)}$.

Thus $\phi_i = \theta_i\psi_i$ is an affine continuous map of S onto K_i such that $\phi_i(e(M_{ij})) = e_{ij}$ for all $j = 1,\dots,n(i)$ and $\phi_i(e(M)) = \phi_i(r) = 0$ for all nonzero idempotent maximal ideals $M \notin X_i$.

Now consider any $i \in I_2$. According to [5, Corollary 21], M_i is contained in a set $\{N_1,\dots,N_k\}$ of distinct nonzero idempotent maximal ideals of R such that $O_r(N_j) = O_\ell(N_{j+1})$ for all $j = 1,\dots,k-1$, while $O_r(N) \neq O_\ell(N_1)$ and $O_r(N_k) \neq O_\ell(N)$ for all nonzero idempotent maximal ideals N of R. By choosing a suitable affine combination of maps of the form $s \mapsto s([I])$, where I is a power of an intersection of a subset of the N_j, we obtain an affine continuous map $\phi_i : S \to K_i$ such that $\phi_i(e(M_i)) = 1$ and $\phi_i(e(M)) = \phi_i(r) = 0$ for all nonzero idempotent maximal ideals $M \neq M_i$.

These maps ϕ_i induce an affine continuous map $\phi : S \to E$. Observing that each $K_i' \subseteq \phi(S)$, we see that $K \subseteq \phi(S)$. On the other hand, with the aid of Theorem 7 we infer that if s is any extreme point of S, then $\phi(s) \in K$. Consequently, it follows from the Krein-Milman Theorem that $\phi(S) = K$. If $s,t \in S$ and $\phi(s) = \phi(t)$, then it follows from the construction of the maps ϕ_i that $s([M]) = t([M])$ for all nonzero idempotent maximal ideals M of R, from which it can be shown that $s = t$.

Thus ϕ is an affine continuous bijection of S onto K. Inasmuch as S and K are compact Hausdorff spaces, ϕ is a homeomorphism. □

For low dimensions, it is easy to read off from Theorem 8 just which compact convex sets can occur for S, as we now do.

DIMENSION 0. For S to be zero-dimensional, it must be a single point. This occurs exactly when R has no nonzero idempotent maximal ideals, i.e., when R is a Dedekind prime ring.

DIMENSION 1. For S to be one-dimensional, it must be affinely homeomorphic to $[0,1]$. This happens in either of two ways.

(a) R has exactly two idempotent maximal ideals M_1 and M_2, and they form a cycle.

In this case we may sketch S as follows:

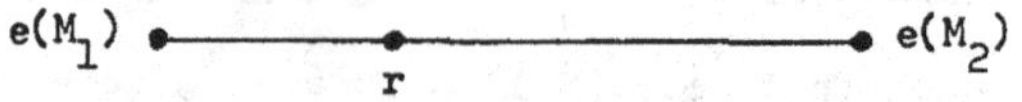

The position of r in S depends on the ratio of $\ell(R/M_1)$ to $\ell(R/M_2)$, as in Proposition 6. For example, if $R = \begin{pmatrix} \mathbb{Z} & 2\mathbb{Z} \\ \mathbb{Z} & \mathbb{Z} \end{pmatrix}$, then $\ell(R/M_1) = \ell(R/M_2) = 1$, in which case r is the midpoint of S. If $R = \begin{pmatrix} \mathbb{Z} & 2\mathbb{Z} & 2\mathbb{Z} \\ \mathbb{Z} & \mathbb{Z} & \mathbb{Z} \\ \mathbb{Z} & \mathbb{Z} & \mathbb{Z} \end{pmatrix}$, then we may label things to that $\ell(R/M_1) = 1$ and $\ell(R/M_2) = 2$, in which case r lies two-thirds of the way from $e(M_1)$ to $e(M_2)$.

(b) R has exactly one nonzero idempotent maximal ideal M (and thus no cycles).

In this case we may sketch S as follows:

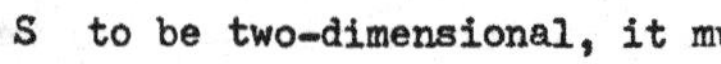

For an example, let T be a simple HNP ring which is not artinian, and let R be the idealizer of a maximal right ideal M of T.

<u>DIMENSION</u> <u>2</u>. For S to be two-dimensional, it must be affinely homeomorphic to either a triangle or a plane quadrilateral. There are three ways for S to be a triangle and one way for S to be a quadrilateral.

(a) R has exactly three idempotent maximal ideals M_1, M_2, M_3, and they form a cycle.

In this case, we may sketch S as follows:

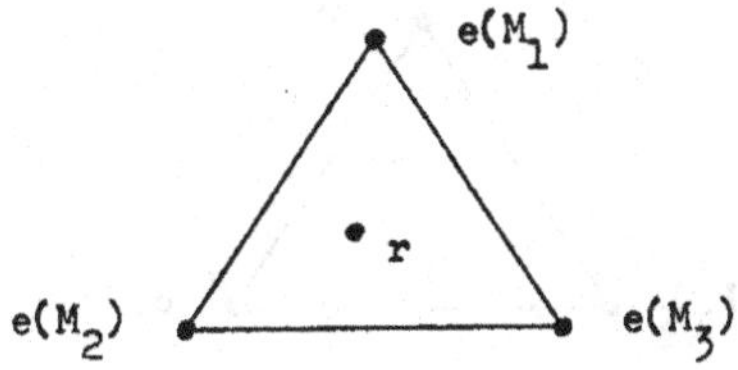

For example, if $R = \begin{pmatrix} \mathbb{Z} & 2\mathbb{Z} & 2\mathbb{Z} \\ \mathbb{Z} & \mathbb{Z} & 2\mathbb{Z} \\ \mathbb{Z} & \mathbb{Z} & \mathbb{Z} \end{pmatrix}$, then each $\ell(R/M_i) = 1$, and r lies at the barycenter of S. If the lengths $\ell(R/M_i)$ are not all equal, as in the case $R = \begin{pmatrix} \mathbb{Z} & 2\mathbb{Z} & 2\mathbb{Z} & 2\mathbb{Z} \\ \mathbb{Z} & \mathbb{Z} & 2\mathbb{Z} & 2\mathbb{Z} \\ \mathbb{Z} & \mathbb{Z} & \mathbb{Z} & \mathbb{Z} \\ \mathbb{Z} & \mathbb{Z} & \mathbb{Z} & \mathbb{Z} \end{pmatrix}$, then r does not lie at the barycenter of S.

(b) R has exactly three idempotent maximal ideals M_1, M_2, M_3, and $\{M_1, M_2\}$ is a cycle.

In this case we may sketch S as follows:

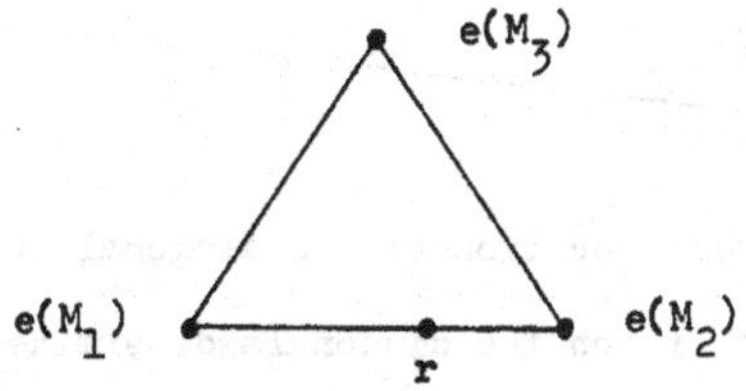

For an example, start with an HNP ring P which has exactly two idempotent maximal ideals M and N but no cycles (such a ring may be constructed as in (c) below), and take $R = \begin{pmatrix} P & M \\ P & P \end{pmatrix}$. In this example $\ell(R/M_1) = \ell(R/M_2) = 1$, hence r lies at the midpoint of an edge of S.

(c) R has exactly two idempotent maximal ideals M_1 and M_2, and they do not form a cycle.

In this case we may sketch S as follows:

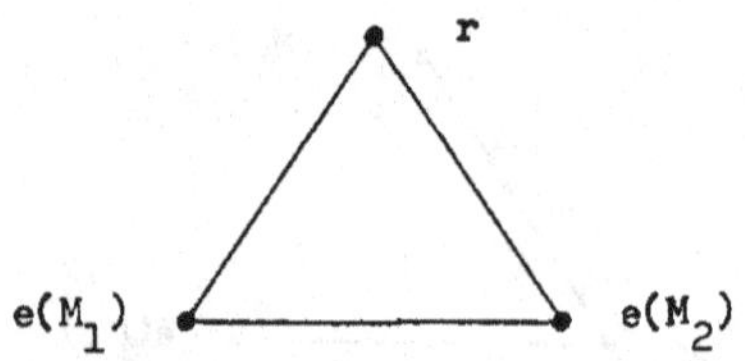

For an example, let T be a simple HNP ring which is not artinian, let K_1 and K_2 be maximal right ideals of T such that $T/K_1 \not\cong T/K_2$, and let R be the idealizer of $K_1 \cap K_2$.

(d) R has exactly four idempotent maximal ideals M_1, M_2, K_1, K_2, and they form cycles $\{M_1, M_2\}$, $\{K_1, K_2\}$.

In this case we may sketch S as follows:

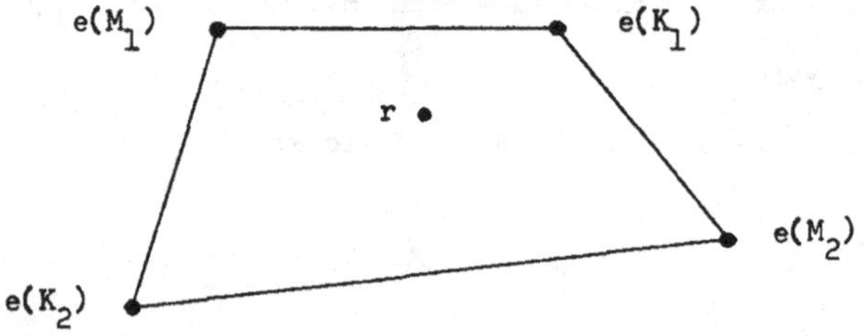

Here r must lie at the intersection of the diagonals of S, and the proportional positions of r on the diagonals determine the shape of S. For example, if $R = \begin{pmatrix} \mathbb{Z} & 6\mathbb{Z} \\ \mathbb{Z} & \mathbb{Z} \end{pmatrix}$, then $\ell(R/M_i) = \ell(R/K_i) = 1$ and so r lies at the midpoint of each diagonal, in which case S is (affinely homeomorphic to) a square, as we saw in the previous section. If $R = \begin{pmatrix} \mathbb{Z} & 6\mathbb{Z} & 6\mathbb{Z} \\ \mathbb{Z} & \mathbb{Z} & \mathbb{Z} \\ \mathbb{Z} & \mathbb{Z} & \mathbb{Z} \end{pmatrix}$, then we may label things so that $\ell(R/M_1) = \ell(R/K_1) = 1$ and $\ell(R/M_2) = \ell(R/K_2) = 2$, in which case S is a trapezoid. If $R = \begin{pmatrix} \mathbb{Z} & 2\mathbb{Z} & 6\mathbb{Z} & 6\mathbb{Z} & 6\mathbb{Z} \\ \mathbb{Z} & \mathbb{Z} & 3\mathbb{Z} & 3\mathbb{Z} & 3\mathbb{Z} \\ \mathbb{Z} & \mathbb{Z} & \mathbb{Z} & \mathbb{Z} & \mathbb{Z} \\ \mathbb{Z} & \mathbb{Z} & \mathbb{Z} & \mathbb{Z} & \mathbb{Z} \\ \mathbb{Z} & \mathbb{Z} & \mathbb{Z} & \mathbb{Z} & \mathbb{Z} \end{pmatrix}$, then we may label things so that $\ell(R/M_1) = 1$, $\ell(R/M_2) = 4$, $\ell(R/K_1) = 2$, $\ell(R/K_2) = 3$, in which case S is not a trapezoid.

<u>DIMENSION</u> 3. For S to be three-dimensional, it must be affinely homeomorphic to a tetrahedron, a quadrilateral-base pyramid, a hexahedron, or an octahedron. There are four ways for S to be a tetrahedron and one way for S to be each of the other shapes.

(a) R has exactly four idempotent maximal ideals, and they form a single cycle.

In this case S is a tetrahedron and r lies in the interior of S. For an example, take $R = \begin{pmatrix} \mathbb{Z} & 2\mathbb{Z} & 2\mathbb{Z} & 2\mathbb{Z} \\ \mathbb{Z} & \mathbb{Z} & 2\mathbb{Z} & 2\mathbb{Z} \\ \mathbb{Z} & \mathbb{Z} & \mathbb{Z} & 2\mathbb{Z} \\ \mathbb{Z} & \mathbb{Z} & \mathbb{Z} & \mathbb{Z} \end{pmatrix}$.

(b) R has exactly four idempotent maximal ideals, of which three form a cycle.

In this case S is a tetrahedron and r lies in the interior of a face of S. An example may be constructed similar to 2(b) above.

(c) R has exactly four idempotent maximal ideals, of which two form a cycle and the other two do not.

In this case S is a tetrahedron and r lies in the interior of an edge of S. An example may be constructed similar to 2(b) above.

(d) R has exactly three idempotent maximal ideals, none of which form cycles.

In this case S is a tetrahedron and r lies at a vertex of S. An example may be constructed similar to 2(c) above.

(e) R has exactly five idempotent maximal ideals, of which four form two cycles of length two.

In this case S is a quadrilateral-base pyramid, and r lies at the intersection of the diagonals of the base of S. An example may be constructed similar to 2(b) above.

(f) R has exactly five idempotent maximal ideals, forming one cycle of length two and one cycle of length three.

In this case S is a hexahedron, and r lies at the intersection of the line determined by e_1, e_2 and the triangle determined by e_3, e_4, e_5, where e_1, e_2 are those vertices of S which are not adjacent to every other vertex and e_3, e_4, e_5 are those vertices of S which are adjacent to every other vertex. For an example, take

$$R = \begin{pmatrix} \mathbb{Z} & \mathbb{Z} & 3\mathbb{Z} & 6\mathbb{Z} & 6\mathbb{Z} & 6\mathbb{Z} \\ \mathbb{Z} & \mathbb{Z} & 3\mathbb{Z} & 6\mathbb{Z} & 6\mathbb{Z} & 6\mathbb{Z} \\ \mathbb{Z} & \mathbb{Z} & \mathbb{Z} & 2\mathbb{Z} & 6\mathbb{Z} & 6\mathbb{Z} \\ \mathbb{Z} & \mathbb{Z} & \mathbb{Z} & \mathbb{Z} & 3\mathbb{Z} & 3\mathbb{Z} \\ \mathbb{Z} & \mathbb{Z} & \mathbb{Z} & \mathbb{Z} & \mathbb{Z} & \mathbb{Z} \\ \mathbb{Z} & \mathbb{Z} & \mathbb{Z} & \mathbb{Z} & \mathbb{Z} & \mathbb{Z} \end{pmatrix}$$

(g) R has exactly six idempotent maximal ideals, forming three cycles of length two.

In this case S is an octahedron and r lies at the intersection of the diagonals of S. For an example, take $R = \begin{pmatrix} \mathbb{Z} & 30\mathbb{Z} \\ \mathbb{Z} & \mathbb{Z} \end{pmatrix}$.

REFERENCES

1. E. M. Alfsen, Compact Convex Sets and Boundary Integrals
   Ergebnisse der Math., Band 57
   Berlin (1971) Springer-Verlag.

2. G. M. Bergman, "Coproducts and some universal ring constructions"
   Trans. American Math. Soc. 200 (1974) 33-88.

3. D. Eisenbud and J. C. Robson, "Hereditary noetherian prime rings"
   J. Algebra 16 (1970) 86-104.

4. K. R. Goodearl, Von Neumann Regular Rings
   London (197-) Pitman.

5. K. R. Goodearl and R. B. Warfield, Jr., "Simple modules over hereditary noetherian prime rings"
J. Algebra (to appear).

6. J. L. Kelley and I. Namioka, Linear Topological Spaces
Princeton (1963) Van Nostrand.

7. T. H. Lenagan, "Bounded hereditary noetherian prime rings"
J. London Math. Soc. 6 (1973) 241-246.

8. R. R. Phelps, Lectures on Choquet's Theorem
Princeton (1966) Van Nostrand.

9. R. G. Swan, Algebraic K-Theory
Springer Lecture Notes No. 76
Berlin (1968) Springer-Verlag.

10. J. C. Robson, "Idealizers and hereditary noetherian prime rings"
J. Algebra 22 (1972) 45-81.

11. Z. Semadeni, Banach Spaces of Continuous Functions
Warsaw (1971) PWN (Polish Scientific Publishers).

University of Utah
Salt Lake City, Utah 84112
U.S.A.

SIMPLE NOETHERIAN RINGS — THE ZALESSKII-NEROSLAVSKII EXAMPLES

K. R. Goodearl

This paper is a community service project which was requested by a group of participants at the conference. Its purpose is to provide a source in English for the details of two important examples constructed by A. E. Zalesskii and O. M. Neroslavskii, which up to now have been generally available only in Russian [7,8]. The first of these examples is a simple noetherian ring which is not isomorphic to a full matrix ring over an integral domain. Our presentation of this example is based on a set of seminar notes by A. W. Chatters [2], with some simplifications worked out in discussions with J. H. Cozzens. The second example is a simple noetherian ring which is not Morita-equivalent to an integral domain. Our presentation of this example is a modification of the presentation in [8], where it is only proved that the ring has no nontrivial idempotents, followed by an application of J. T. Stafford's result that this lack of nontrivial idempotents is sufficient to imply the lack of Morita-equivalence to an integral domain [5].

We shall abbreviate two-sided hypotheses by omitting the phrase "left and right". In particular, we use "noetherian ring" to mean "left and right noetherian ring". Since both of the examples are presented as skew group rings, we begin by recalling the definition and a few basic properties.

Suppose that G is a group of automorphisms of a ring S, with the convention that the action of a group element $g \in G$ on a ring element $s \in S$ is denoted s^g. The skew group ring of G over S is the universal ring extension of S in which G extends to a group of inner automorphisms. Specifically, this skew group ring R is additively a free right S-module with G as basis, and multiplication is defined so that $sg = gs^g$ for all $s \in S$ and $g \in G$ (so that $s^g = g^{-1}sg$). Note that when g has order 2, then we have $gs = s^g g$ as well.

If S is noetherian and G is reasonably small, then R is noetherian as well. For example, if G is finite, then R is finitely generated as an S-module on either side, whence R is noetherian as an S-module on either side, and so R is a noetherian ring. Also, if G is infinite cyclic, then R may be viewed as a localization of a skew polynomial ring $S[x,g]$ (where g is a generator for G), and it follows from the noetherianness of $S[x,g]$ that R is noetherian.

PROPOSITION 1. Let G be a group of automorphisms of a simple ring S, and let R be the skew group ring of G over S. If the identity is the only element of G which is an inner automorphism of S, then R is a simple ring.

Proof. [4, Proposition 1.1]. □

PROPOSITION 2. Let G be a finite group of automorphisms of a hereditary ring S, and let R be the skew group ring of G over S. If R is a simple ring and the order of G is invertible in S, then R is a hereditary ring.

Proof. There is a natural right R-module structure on S, where the module multiplication $*$ is defined so that $s*(gt) = s^g t$ for all $s,t \in S$ and $g \in G$. It is easy to ckeck, as in [3, Lemma 1.2, Corollary 1.4] that S is a cyclic projective right R-module, and that $\mathrm{End}_R(S_R)$ is isomorphic to the fixed ring $S^G = \{s \in S \mid s^g = s$ for all $g \in G\}$. Since R is a simple ring, S_R is a generator, whence R is Morita-equivalent to S^G. According to [1, Propositions 1.1, 1.2], S^G is a hereditary ring, hence so is R. □

EXAMPLE I. There exists a simple noetherian hereditary ring R which is not isomorphic to a full matrix ring over an integral domain.

Proof. Let F be any subfield of the real numbers, and let S be the Weyl algebra $A_1(F)$, that is, the F-algebra with generators x and θ subject to the sole relation $\theta x - x\theta = 1$. It is well-known that S is a simple, noetherian, hereditary integral domain.

Let h be the F-automorphism of S such that $x^h = -x$ and $\theta^h = -\theta$, and let R be the skew group ring of the two element group {1,h} over S. Note that R is noetherian. Observing that all invertible elements of S lie in F and thus are central, we see that h is not an inner automorphism of S. Thus R is simple, by Proposition 1. Since $1/2 \in S$, Proposition 2 now says that R is hereditary.

Observing that $(1+h)(1-h) = 0$, we see that R is not an integral domain, hence the uniform dimension of R_R is at least 2. On the other hand, R_S is a free module of rank 2 and so has uniform dimension 2. Thus the uniform dimension of R_R is 2 as well. Consequently, if $R \cong M_n(D)$ for some integral domain D, then $n = 2$. In this case, there is an element $r \in R$ corresponding to the matrix $\begin{pmatrix} 0 & 1 \\ -1 & 0 \end{pmatrix}$, so that $r^2 = -1$. We show that no such element r exists in R.

Write $r = a + bh$ for some $a,b \in S$. Expanding the equation $r^2 = -1$ and comparing coefficients, we obtain

(1) $$a^2 + bb^h = -1$$

(2) $$ab + ba^h = 0.$$

Since all the invertible elements of S lie in F, we see that it is not possible to have either $a^2 = -1$ or $bb^h = -1$, hence it follows from (1) that

$a \neq 0$ and $b \neq 0$.

For any nonzero $p \in S$, we use $\deg_\theta(p)$ to denote the degree of p considered as a polynomial in θ with coefficients from $F[x]$, and we use $\deg_x(p)$ to denote the degree of p considered as a polynomial in x with coefficients from $F[\theta]$. We claim that $\deg_\theta(a)$ and $\deg_x(a)$ are both even.

Write

(3) $$a = a_0 + a_1\theta + \dots + a_n\theta^n$$

(4) $$b = b_0 + b_1\theta + \dots + b_k\theta^k$$

for some $a_i, b_j \in F[x]$ such that $a_n \neq 0$ and $b_k \neq 0$, so that $n = \deg_\theta(a)$. If $n > k$, then comparing coefficients of θ^{2n} in (1) we obtain $a_n^2 = 0$, which is false. Likewise, if $k > n$ then we obtain $b_k b_k^h = 0$, which is also false. Thus $k = n$. If $n = 0$, then a_0 and b_0 are polynomials in $F[x]$ such that $a_0^2 + b_0 b_0^h = -1$. Taking constant terms, we obtain $\alpha_0, \beta_0 \in F$ such that $\alpha_0^2 + \beta_0^2 = -1$, which is impossible. Therefore $n = k > 0$.

Let T denote the ring $F(x)[\theta]$, which is an over-ring of S. Inasmuch as $ab = -ba^h$ by (2), we see that left multiplication by a induces a right T-module endomorphism φ of T/bT. Since $b_n = b_k$ is invertible in $F(x)$, we infer that T/bT is an n-dimensional right vector space over $F(x)$ (with basis $\bar{1}, \bar{\theta}, \dots, \bar{\theta}^{n-1}$). Using (1), we see that $\varphi^2 = -1$, whence $(\det \varphi)^2 = (-1)^n$. Thus $(-1)^n$ has a square root in $F(x)$, which happens only when n is even.

Therefore $\deg_\theta(a)$ is even, as claimed. Inasmuch as there is an F-automorphism g of S such that $x^g = \theta$ and $\theta^g = x$, and also $hg = gh$, we infer that $\deg_x(a)$ is even as well.

Since $n > 0$, comparing coefficients of θ^{2n} in (1) shows that

(5) $$a_n^2 + b_n b_n^h = 0 .$$

Write

(6) $$a_n = \alpha_0 + \alpha_1 x + \dots + \alpha_m x^m$$

(7) $$b_n = \beta_0 + \beta_1 x + \dots + \beta_\ell x^\ell$$

for some $\alpha_i, \beta_j \in F$ such that $\alpha_m \neq 0$ and $\beta_\ell \neq 0$. If $m > \ell$, then comparing coefficients of x^{2m} in (5) shows that $\alpha_m^2 = 0$, which is false. Likewise, if $\ell > m$ then we obtain $\beta_\ell^2 = 0$, which is also false. Thus $m = \ell$. Comparing coefficients of x^{2m} in (5) once more, we find that $\alpha_m^2 + (-1)^m \beta_m^2 = 0$. Since $\alpha_m \neq 0$ and $\beta_m \neq 0$, it follows that m must be odd.

Now choose an odd integer $t > \deg_x(a)$. There is an F-automorphism f of S such that $x^f = x$ and $\theta^f = \theta + x^t$. Note that $x^{hf} = -x = x^{fh}$ and $\theta^{hf} = -\theta - x^t = \theta^{fh}$ (because t is odd), whence $fh = hf$. Setting $c = a^f$ and $d = b^f$ and applying f to (1) and (2), we obtain $c^2 + dd^h = -1$ and $cd + dc^h = 0$. Consequently, the argument above shows that $\deg_x(c)$ is even.

Write $a = \sum_{(i,j)\in K} \gamma_{ij} x^i \theta^j$, where K is a nonempty finite set of ordered pairs of nonnegative integers and each γ_{ij} is a nonzero element of F. Then

(8) $$c = a^f = \sum_{(i,j)\in K} \gamma_{ij} x^i (\theta + x^t)^j .$$

To find the partial sum involving only indices of the form (i,n), we use (3) and (6). It follows that this partial sum equals $(\alpha_0 + \alpha_1 x + \dots + \alpha_m x^m)(\theta + x^t)^n$, in which the highest power of x is x^{m+tn}. Whenever $(i,j) \in K$ and $j < n$, we have

$$i + tj \leq \deg_x(a) + tj < t + tj \leq tn \leq m + tn ,$$

hence $\gamma_{ij} x^i (\theta + x^t)^j$ is of degree at most $m + tn - 1$ in x. As a result, we conclude that $\deg_x(c) = m + tn$. Inasmuch as m is odd and $n = \deg_x(a)$ is even, this contradicts the fact that $\deg_x(c)$ is even.

Therefore there does not exist $r \in R$ such that $r^2 = -1$. □

In skew group rings of the two element group, there is an easy situation in which 2×2 matrix units occur. Namely, let S be a ring containing $1/2$, let h be an automorphism of S of period 2, and let R be the skew group ring of $\{1,h\}$ over S. Then $e_{11} = (1+h)/2$ and $e_{22} = (1-h)/2$ are orthogonal idempotents in R and $e_{11}+e_{22} = 1$. If there exists a unit $u \in S$ such that $u^h = -u$, set $e_{12} = ue_{22}$ and $e_{21} = u^{-1}e_{11}$. In this case, it is easily checked that $\{e_{11},e_{12},e_{21},e_{22}\}$ is a set of 2×2 matrix units in R. If this is the only way in which R can contain 2×2 matrix units (assuming S does not), then it would become trivial to check that Example I and similar examples are not matrix rings over domains. In the hope that this is indeed the case, we make the following conjecture.

<u>CONJECTURE I</u>. Let S be a simple noetherian integral domain which contains $1/2$, let h be an automorphism of S of period 2, and let R be the skew group ring of $\{1,h\}$ over S. Then $R \cong M_2(D)$ for some integral domain D if and only if there exists a unit $u \in S$ such that $u^h = -u$. □

<u>PROPOSITION 3</u>. Let R be a simple noetherian ring of Krull dimension 1. If R is Morita-equivalent to an integral domain, then either R is an integral domain or R contains a nontrivial idempotent.

<u>Proof</u>. Use the proof of [5, Theorem 3]. □

<u>EXAMPLE II</u>. There exists a simple noetherian ring R of Krull dimension 1 which is not Morita-equivalent to an integral domain.

<u>Proof</u>. Let F be a field of characteristic 2, let z and a be independent indeterminates, and set $T = F(z)[a,a^{-1}]$. Note that T is a commutative noetherian domain. Let b be the $F(z)$-automorphism of T such

that $a^b = z^{-1}a$, and let S be the skew group ring over T of the infinite cyclic group generated by b. We note that S is a noetherian integral domain, and we use Proposition 1 to see that S is simple. In S, note that $ab = ba^b = bz^{-1}a$, whence $ba = zab$.

Let h be the $F(z)$-automorphism of S such that $a^h = a^{-1}$ and $b^h = b^{-1}$, and let R be the skew group ring of $\{1,h\}$ over S. Note that R is noetherian, and use Proposition 1 to see that R is simple. According to [5, Theorem 2], R has Krull dimension 1. Note that $(1+h)(1-h) = 0$, so that R is not an integral domain. In order to prove that R is not Morita-equivalent to an integral domain, Proposition 3 says that it suffices to show that R contains no nontrivial idempotents.

Suppose that e is a nontrivial idempotent in R. Writing $e = f + gh$ for some $f, g \in S$ and calculating the relation $e = e^2$, we obtain

(1) $$f^2 + gg^h = f$$

(2) $$fg + gf^h = g\,.$$

Since S is an integral domain, it contains no nontrivial idempotents. Thus $e \notin S$, whence $g \neq 0$. Then $gg^h \neq 0$, hence we see from (1) that $f^2 \neq f$. Thus $f \neq 0$ and $f \neq 1$.

We may expand f and g as finite sums

(3) $$f = \Sigma f_{ij}a^ib^j \qquad ; \qquad g = \Sigma g_{ij}a^ib^j$$

for some $f_{ij}, g_{ij} \in F(z)$, indexed by a subset of $Z \times Z$. Order $Z \times Z$ lexicographically, and let (n,k) be the largest element of $Z \times Z$ for which $g_{nk} \neq 0$.

For $(s,t) \geq (0,0)$, we compare coefficients for $a^{n+s}b^{k+t}$ in (2), and we obtain

(4) $$g_{n+s,k+t} = \sum_{i,j} g_{n-i,k-j}[f_{i+s,j+t}z^{(j+t)(n-i)} + f_{-i-s,-j-t}z^{(k-j)(i+s)}]\,.$$

Since $g_{n-i,k-j} = 0$ for $(i,j) < (0,0)$, the summation in (4) is taken over all $(i,j) \geq (0,0)$. Moving the $(0,0)$ term in (4) to the left-hand side, we obtain

$$(5)\quad g_{n+s,k+t} + g_{n,k}(f_{s,t}z^{tn} + f_{-s,-t}z^{ks}) = \sum_{(i,j)>(0,0)} g_{n-i,k-j}[f_{i+s,j+t}z^{(j+t)(n-i)} + f_{-i-s,-j-t}z^{(k-j)(i+s)}].$$

The case $(s,t) = (0,0)$ of (5) simplifies as follows, where we have replaced i and j by i_1 and j_1. (This is the only place in the proof where characteristic 2 is needed.)

$$(6)\quad g_{n,k} = \sum_{(i_1,j_1)>(0,0)} g_{n-i_1,k-j_1}[f_{i_1,j_1}z^{nj_1} + f_{-i_1,-j_1}z^{ki_1}]z^{-i_1j_1}.$$

For the case $(s,t) > (0,0)$, we have

$$(7)\quad g_{n,k}(f_{s,t}z^{tn} + f_{-s,-t}z^{ks}) = \sum_{(i,j)>(0,0)} g_{n-i,k-j}[f_{i+s,j+t}z^{n(j+t)}z^{-i(j+t)} + f_{-i-s,-j-t}z^{k(i+s)}z^{-j(i+s)}].$$

In order to obtain a contradiction, we shall generalize (6) to the following formula (by induction on m):

$$(8)\quad g_{n,k}^m = \sum g_{n-i_1,k-j_1}g_{n-i_2,k-j_2}\cdots g_{n-i_m,k-j_m} \cdot [f_{i_1+\cdots+i_m,j_1+\cdots+j_m}z^{n(j_1+\cdots+j_m)} + f_{-i_1-\cdots-i_m,-j_1-\cdots-j_m}z^{k(i_1+\cdots+i_m)}] \cdot z^{-i_1j_1-i_2(j_1+j_2)-\cdots-i_m(j_1+\cdots+j_m)}.$$

The summation in (8) is taken over all possible admissible pairs of integers $(i_1,j_1) > (0,0), \ldots, (i_m,j_m) > (0,0)$.

To see that (8) is enough to reach a contradiction, we check that for suitably large m, each term in (8) is zero. Namely, choose a positive integer q such that $f_{u,v} = 0$ whenever $|u| > q$ or $|v| > q$, and choose a negative integer r such that $g_{n-i,k-j} = 0$ whenever $i \geqq 0$ and $j \leqq r$. Setting $m = q(2-r)$, we infer that for each term in (8) we must have $|i_1+\dots+i_m| > q$ or $|j_1+\dots+j_m| > q$ or some $j_\alpha \leqq r$. Thus each term in (8) is indeed zero, whence $g_{n,k}^m = 0$. But then $g_{n,k} = 0$, which is false.

Thus it only remains to establish (8). The case $m = 1$ is (6). Now assume that (8) holds for a given m. Multiplying (8) by $g_{n,k}$ and utilizing (7), we obtain

$$(9)\quad g_{n,k}^{m+1} = \Sigma\, g_{n-i_1,k-j_1} g_{n-i_2,k-j_2} \cdots g_{n-i_{m+1},k-j_{m+1}} \cdot$$

$$\cdot \{f_{i_1+\dots+i_{m+1},j_1+\dots+j_{m+1}} z^{n(j_1+\dots+j_{m+1})} z^{-i_{m+1}(j_1+\dots+j_{m+1})} +$$

$$f_{-i_1-\dots-i_{m+1},-j_1-\dots-j_{m+1}} z^{k(i_1+\dots+i_{m+1})} z^{-j_{m+1}(i_1+\dots+i_{m+1})}\}.$$

$$\cdot z^{-i_1j_1-i_2(j_1+j_2)-\dots-i_m(j_1+\dots+j_m)} .$$

This summation is taken over all possible $(i_1,j_1) > (0,0)$, ..., $(i_{m+1},j_{m+1}) > (0,0)$.

Now group the terms in (9) corresponding to $(2m+2)$-tuples $(i_1,j_1,\dots,i_{m+1},j_{m+1})$ lying in one orbit relative to the natural action of the permutation group S_{m+1} (that is, S_{m+1} acts by permuting the subscripts). Let H_{m+1} be the stabilizer of $(i_1,j_1,\dots,i_{m+1},j_{m+1})$ in S_{m+1}, and let T_{m+1} be a transversal of H_{m+1} in S_{m+1}, that is, a fixed set of representatives of the distinct cosets sH_{m+1}. Note that this is ambiguous notation, since H_{m+1} and T_{m+1} depend on which orbit is under scrutiny. The result of this grouping is an expression of the form

$$(10)\quad g^{m+1}_{n,k} = \Sigma g_{n-i_1,k-j_1} g_{n-i_2,k-j_2} \cdots g_{n-i_{m+1},k-j_{m+1}} \cdot$$

$$\cdot (f_{i_1+\dots+i_{m+1},j_1+\dots+j_{m+1}} z^{n(j_1+\dots+j_{m+1})} \cdot$$

$$\cdot [\sum_{\sigma \in T_{m+1}} z^{\sigma[-i_1 j_1 - \dots - i_{m+1}(j_1+\dots+j_{m+1})]}]$$

$$+ f_{-i_1-\dots-i_{m+1},-j_1-\dots-j_{m+1}} z^{k(i_1+\dots+i_{m+1})} \cdot$$

$$\cdot [\sum_{\sigma \in T_{m+1}} z^{\sigma[-i_1 j_1 - \dots - i_m(j_1+\dots+j_m) - j_{m+1}(i_1+\dots+i_{m+1})]}]) .$$

Here the outside summation is taken over the representatives of the orbits of S_{m+1} on the set

$$\{(i_1,j_1,\dots,i_{m+1},j_{m+1}) \mid (i_1,j_1) > (0,0), \dots, (i_{m+1},j_{m+1}) > (0,0)\} .$$

Also, we are using $\sigma[-i_1 j_1 - \dots]$ to denote $-i_{\sigma(1)} j_{\sigma(1)} - \dots$. In order to show that (10) yields the desired equation, we shall prove that the summations in square brackets in (10) are equal, so that they may be factored out.

First note that

$$(11)\quad i_1 j_1 + i_2(j_1+j_2) + \dots + i_{m+1}(j_1 + \dots + j_{m+1})$$
$$= j_{m+1} i_{m+1} + j_m(i_{m+1} + i_m) + \dots + j_1(i_{m+1} + \dots + i_1) .$$

Setting $t = \begin{pmatrix} 1 & 2 & \dots & m & m+1 \\ m+1 & m & \dots & 2 & 1 \end{pmatrix} \in S_{m+1}$ and using (11), we see that

$$(12)\quad \sum_{\sigma \in T_{m+1}} z^{\sigma[-i_1 j_1 - i_2(j_1+j_2) - \dots - i_{m+1}(j_1+\dots+j_{m+1})]}$$

$$= \sum_{\sigma \in T_{m+1}} z^{\sigma t[-i_1 j_1 - i_2(j_1+j_2) - \dots - i_{m+1}(j_1+\dots+j_{m+1})]}$$

$$= \sum_{\sigma \in T_{m+1}} z^{\sigma[-j_1 i_1 - j_2(i_1+i_2) - \dots - j_{m+1}(i_1+\dots+i_{m+1})]} .$$

Now identify S_m with the subgroup of S_{m+1} consisting of those permutations which fix $m+1$. The orbit of $(i_1,j_1,\ldots,i_{m+1},j_{m+1})$ relative to S_{m+1} is a disjoint union of orbits relative to S_m. Using H_m and T_m as above, we infer from (12) that the following equation holds for each S_m-orbit of $(i_1,j_1,\ldots,i_{m+1},j_{m+1})$:

$$(13)\qquad \sum_{\sigma\in T_m} z^{\sigma[-i_1j_1-\cdots-i_m(j_1+\cdots+j_m)-j_{m+1}(i_1+\cdots+i_{m+1})]} = \sum_{\sigma\in T_m} z^{\sigma[-j_1i_1-\cdots-j_m(i_1+\cdots+i_m)-j_{m+1}(i_1+\cdots+i_{m+1})]} .$$

Adding the equations (13) over each S_m-orbit contained in some S_{m+1}-orbit, we obtain

$$(14)\qquad \sum_{\sigma\in T_{m+1}} z^{\sigma[-i_1j_1-\cdots-i_m(j_1+\cdots+j_m)-j_{m+1}(i_1+\cdots+i_{m+1})]} = \sum_{\sigma\in T_{m+1}} z^{\sigma[-j_1i_1-\cdots-j_{m+1}(i_1+\cdots+i_{m+1})]}$$

for each S_{m+1}-orbit of $(i_1,j_1,\ldots,i_{m+1},j_{m+1})$. Combining (14) with (12) yields

$$(15)\qquad \sum_{\sigma\in T_{m+1}} z^{\sigma[-i_1j_1-\cdots-i_m(j_1+\cdots+j_m)-j_{m+1}(i_1+\cdots+i_{m+1})]} = \sum_{\sigma\in T_{m+1}} z^{\sigma[-i_1j_1-i_2(j_1+j_2)-\cdots-i_{m+1}(j_1+\cdots+j_{m+1})]} ,$$

as desired.

Thus the summations in square brackets in (10) are indeed equal, hence the expression $\sum_{\sigma\in T_{m+1}} z^{\sigma[-i_1j_1-\cdots-i_{m+1}(j_1+\cdots+j_{m+1})]}$ may be carried outside the braces in (10). With the aid of this, (9) may be rearranged to complete the induction step, i.e., to show that (8) is valid with m replaced by $m+1$.

Thus the induction works, so that (8) holds for all m, implying the desired contradiction. Therefore R contains no nontrivial idempotents. □

As noted in [5], the ring R constructed in Example II has infinite global dimension. The situation is nicer in the case of finite global dimension, by [6, Corollary 3]: Every simple noetherian ring with finite global dimension and Krull dimension 1 is Morita-equivalent to an integral domain. This lends some plausibility to the following question.

<u>QUESTION</u>. Is every simple noetherian ring of finite global dimension Morita-equivalent to an integral domain? □

In skew group rings of the two element group, there is an easy situation in which nontrivial idempotents occur. Namely, let S be a ring with an automorphism h of period 2, and let R be the skew group ring of $\{1,h\}$ over S. If there exists an element $u \in S$ such that $u + u^h = 1$, then it is easily checked that $u(1+h)$ is a nontrivial idempotent in R. For example, if $1/2 \in S$, then $u = 1/2$ has this property. If this is the only way in which R can contain nontrivial idempotents (assuming S does not), then the proof of Example II could be simplified. In the hope that this is indeed the case, we make the following conjecture.

<u>CONJECTURE II</u>. Let S be a simple noetherian integral domain, let h be an automorphism of S of period 2, and let R be the skew group ring of $\{1,h\}$ over S. Then R contains a nontrivial idempotent if and only if there exists $u \in S$ such that $u + u^h = 1$. □

REFERENCES

1. G. M. Bergman, "Groups acting on hereditary rings"
Proc. London Math. Soc. 23 (1971) 70-82.

2. A. W. Chatters, "The Zalesskii-Neroslavskii example of a simple noetherian ring which is not a matrix ring over an integral domain"
Seminar notes (typescript) (1976).

3. J. W. Fisher and J. Osterburg, "Some results on rings with finite group actions"
(to appear).

4. D. Handelman, J. Lawrence, and W. Schelter, "Skew group rings"
Houston J. Math. (to appear).

5. J. T. Stafford, "A simple noetherian ring not Morita equivalent to a domain"
Proc. American Math. Soc. 68 (1978) 159-160.

6. ——, "Morita equivalence of simple noetherian rings"
(to appear).

7. A. E. Zalesskii and O. M. Neroslavskii, "On simple noetherian rings" (Russian)
Isvestija Akad. Nauk. BSSR 5 (1975) 38-42.

8. ——, "There exists a simple noetherian ring with divisors of zero, but without idempotents" (Russian with English abstract)
Communic. in Algebra 5 (1977) 231-244.

University of Utah

Salt Lake City, Utah 84112

U.S.A.

ANNEAUX DE POLYNOMES SEMI-HEREDITAIRES

par J.M. GOURSAUD et J.L. PASCAUD

Université de Poitiers
40, Avenue du Recteur Pineau
86022 - POITIERS

On se propose d'étudier le problème suivant : étant donnés un anneau unitaire A et un automorphisme σ de A , à quelles conditions l'anneau des polynômes $A[X,\sigma]$ "tordu" par la multiplication $Xa = \sigma(a)X$, est-il semi-héréditaire ?

Dans le cas où A est commutatif et σ trivial, J.P. Soublin [7] et P.J. Mac Carthy [5] ont montré que $A[X]$ est semi-héréditaire si et seulement si A est régulier. Dans [3] on montre que ce résultat ne subsiste pas si A n'est pas commutatif : si A est régulier auto-injectif à gauche (ou à droite), $A[X]$ est semi-héréditaire si et seulement si A est un produit fini d'anneaux de matrices sur des anneaux réduits auto-injectifs.

Dans une première partie on propose quelques caractérisations utiles des anneaux de polynômes semi-héréditaires. Dans la seconde partie on introduit une décomposition par rapport à σ des anneaux réguliers réduits auto-injectifs qui parait adaptée au problème et s'inspire de la décomposition des anneaux réguliers auto-injectifs en types [6] . Enfin dans la dernière partie outre des conditions nécessaires ou suffisantes pour que $A[X,\sigma]$ soit semi-héréditaire, on donne une caractérisation complète lorsque A est produit d'anneaux simples.

Rappels et notations.

a) Un anneau A est de <u>Rickart à gauche</u> si tout idéal à gauche monogène est projectif. A est semi-héréditaire à gauche si et seulement si quel que soit n l'anneau des matrices $M_n(A)$ est de Rickart à gauche (voir [4]).

b) Pour les définitions et propriétés des anneaux réguliers auto-injectifs, on se reportera à [6] .

c) On notera $\ell_{A[X,\sigma]}(P)$ (resp. $r_{A[X,\sigma]}(P)$) l'annulateur à gauche (resp. à droite) de P dans $A[X,\sigma]$. Pour tout idéal à gauche I de $A[X,\sigma]$ on définit une suite d'idéaux à gauche $(c_n(I))_{n\in\mathbb{N}}$ de A par :

$$c_n(I) = \{a\in A \mid \exists Q\in A[X,\sigma] \text{ , degré } (Q)\leq n \text{ , } Q(o) = a\} .$$

On a l'inclusion $c_n(I)\subset C_{n+1}(I)$.

I.- ANNEAUX DE POLYNOMES TORDUS SEMI-HEREDITAIRES A GAUCHE.

LEMME 1.- *Soit* A *un anneau unitaire,* σ *un automorphisme de* A *alors si* $A[X,\sigma]$ *est semi-héréditaire à gauche,* A *est un anneau régulier.*

L'idéal $I = A[X,\sigma]X + A[X,\sigma]x$ étant projectif pour tout élément x de A , il existe des formes linéaires ϕ_0 et ϕ_1 de I dans $A[X,\sigma]$ telles que :

$$Xx = \phi_0(Xx)X + \phi_1(Xx)x$$

$$\sigma(x)X = X\phi_0(x)X + \sigma(x)\phi_1(X)x$$

Il existe donc un élément a de A tel que : $\sigma(x) = \sigma(x)a\sigma(x)$.

σ étant un automorphisme, A est régulier ∎

Désormais, A désignera un anneau régulier. Le lemme suivant est une adaptation de J.B. Castillon.

LEMME 2.- *Soit* A *un anneau régulier,* σ *un automorphisme de* A . *Alors tout idéal à gauche (ou à droite) de* $A[X,\sigma]$ *est plat (i.e.* $\mathrm{WdimA}[X,\sigma]\leq 1$*).*

Soit M un $A[X,\sigma]$-module à gauche. La suite de $A[X,\sigma]$-modules

$$0 \to \ker\phi \to A[X,\sigma] \otimes_A M \overset{\phi}{\to} M \to 0 ,$$

où ϕ est définie par $\phi(\Sigma X^i \otimes m_i) = \Sigma X^i m_i$, est exacte. D'autre part l'application

$$A[X,\sigma] \otimes_A M \overset{\psi}{\to} \ker \phi$$

définie par $\psi(\sum_{i=o}^{n} X^i \otimes m_i) = \sum_{i=o}^{n} X^i \otimes Xm_i - \sum_{i=o}^{n} X^{i+1} \otimes m_i$ est une application $\mathbb{Z}$-linéaire bijective vérifiant : $\psi(ax) = \sigma(a)\psi(x)$.

M étant un A-module plat, les modules $A[X,\sigma] \otimes_A M$ et $(\ker \phi, *)$ sont des $A[X,\sigma]$-modules plats (si $x \in \ker \phi$ et $a \in A$ on définit : $a * x = \sigma(a)x$) , ce qui montre que $\ker \phi$ est un $A[X,\sigma]$-module plat ■

En conséquence du lemme 2 , $A[X,\sigma]$ est semi-héréditaire à gauche si et seulement si, pour tout entier n , l'annulateur à gauche d'un élément P de $M_n(A)[X,\sigma]$ est un idéal de type fini : en effet, dans ce cas, l'idéal à gauche engendré par P est plat et de présentation finie donc projectif. Les deux lemmes suivants caractérisent un tel élément P .

<u>LEMME 3</u>.- *Soit* A *un anneau régulier,* σ *un automorphisme de* A . *Si* P *désigne un élément de* $A[X,\sigma]$ *et* I *l'annulateur* $\ell_{A[X,\sigma]}(P)$ *, les idéaux* $c_n(I)$ *sont de type fini.*

Soit $P = a_m X^m + \ldots + a_o$ et $Q = b_n X^n + \ldots + b_o$, tel que $Q \in I = \ell_{A[X,\sigma]}(P)$

La relation $QP = o$ est équivalente au système :

$$(1)\quad \begin{cases} b_n \sigma^n(a_m) = 0 \\ b_n \sigma^n(a_{m-1}) + b_{n-1}\sigma^{n-1}(a_m) = 0 \\ \ldots\ldots\ldots\ldots\ldots \\ b_n \sigma^n(a_{m-k}) + b_{n-1}\sigma^{n-1}(a_{m-k+1}) + \ldots + b_{n-k}\sigma^{n-k}(a_m) = 0 \\ \ldots\ldots\ldots\ldots\ldots\ldots\ldots\ldots\ldots \\ \qquad\qquad b_1\sigma(a_o) + b_o a_1 = 0 \\ \qquad\qquad b_o a_o = 0 \end{cases}$$

A chaque (n+1)-uplet $(\alpha_o,\ldots,\alpha_n)$ d'éléments de A on associe un sous-A-module de A^{n+1} défini par :

$$L_{n+1}(\alpha_o,\ldots,\alpha_n) = \{(\beta_o,\ldots,\beta_n) \mid \sum_{i=o}^{n} \beta_i\alpha_i = o\}$$

A étant régulier, c'est un module de type fini. On pose :

$$\mathcal{L}_{n+1}(a_o,\dots,a_m) = L_{n+1}(a_o,0,\dots,0) \cap L_{n+1}(a_1,\sigma(a_o),0,\dots,0) \cap \dots$$
$$\dots \cap L_{n+1}(a_i,\sigma(a_{i-1}),\sigma^2(a_{i-2}),\dots,\sigma^i(a_o),0,\dots,0) \cap \dots$$
$$\dots \cap L_{n+1}(0,\dots,0,\sigma^{n-k}(a_m),\dots,\sigma^n(a_{m-k})) \cap \dots$$
$$\dots \cap L_{n+1}(0,\dots,0,\sigma^n(a_m))$$

D'après (1) on a

$$Q \in I \iff (b_o,\dots,b_n) \in \mathcal{L}_{n+1}(a_o,\dots,a_m)$$

$\mathcal{L}_{n+1}(a_o,\dots,a_m)$ étant intersection finie de sous-modules de type fini de A^{n+1}, est un A-module de type fini; son image par la première projection de A^{n+1} est $C_n(I)$ qui est donc un idéal de type fini ∎

<u>LEMME 4</u>.- *Soient* A *un anneau régulier,* σ *un automorphisme de* A. *Soient* P *un élément de* $A[X,\sigma]$ *et* $I = \ell_{A[X,\sigma]}(P)$. *Alors* I *est un idéal de type fini si, et seulement si, la suite* $(C_n(I))$, $n \in \mathbb{N}$ *est stationnaire.*

Si pour tout $k \in \mathbb{N}$ on a $C_N(I) = C_{N+k}(I)$, soit a_i $(o \leq i \leq N)$ un générateur de $C_i(I)$ et Q_i un élément de I tel que $Q_i(o) = a_i$. X n'étant pas diviseur de zéro, on vérifie aisément que $Q_o,\dots,Q_N$ engendrent I ∎

X n'est pas diviseur de zéro et vérifie la condition de Ore. On désigne par $A[X,X^{-1},\sigma]$ le localisé de $A[X,\sigma]$ par rapport à l'ensemble $\{X^n \ ; \ n \in N\}$.

<u>PROPOSITION 1.</u>- *Soit* A *un anneau régulier,* σ *un automorphisme de* A. *Alors les assertions suivantes sont équivalentes :*

a) $A[X,\sigma]$ *est semi-héréditaire à gauche.*

b) Il existe un entier $n \neq 0$, *tel que* $A[X,\sigma^n]$ *soit semi-héréditaire à gauche.*

c) Pour tout entier relatif $n \neq 0$ $A[X,\sigma^n]$ *est semi-héréditaire à gauche.*

d) $A[X,X^{-1},\sigma]$ *est semi-héréditaire à gauche.*

e) $A[X,\sigma^{-1}]$ *est semi-héréditaire à gauche.*

Remarquons d'abord que si n est un entier strictement positif, $A[X,\sigma^n]$ est canoniquement isomorphe au sous-anneau $A[X^n,\sigma]$ de $A[X,\sigma]$. D'autre part l'anneau $A[X,X^{-1},\sigma]$ est canoniquement isomorphe à l'anneau $A[X,X^{-1},\sigma^{-1}]$. Il suffit donc de prouver d) $\Rightarrow$ a) $\Rightarrow$ c) et b) $\Rightarrow$ a).

d) $\Rightarrow$ a) Il suffit de montrer que l'annulateur à gauche I d'un polynôme P de $A[X,\sigma]$ est de type fini. Par hypothèse $\ell_{A[X,X^{-1},\sigma]}(P)$ est engendré par un idempotent

$$E = a_m X^m + \ldots + a_o + a_{-1}X^{-1} + \ldots + a_{-n}X^{-n}$$

Il en résulte :

$$\ell_{A[X,\sigma]}(P) = \{QE \ , \ Q \in A[X,\sigma] \ \text{ et } \ QE \in A[X,\sigma]\}$$

Soit Q un élément de $A[X,\sigma]$. Alors si $Q = Q_1 X^n + b_{n-1}X^{n-1} \ldots + b_o$, QE appartient à $A[X,\sigma]$ si, et seulement si, les relations suivantes sont vérifiées :

$$(2) \quad \begin{cases} 0 = b_o a_{-n} \\ 0 = b_o a_{-(n-1)} + b_1 \sigma(a_{-n}) \\ \ldots\ldots\ldots\ldots \\ 0 = b_o a_{-1} + b_1\sigma(a_{-2}) + \ldots + b_{n-1}\sigma^{n-1}(a_{-n}) \end{cases}$$

Or l'ensemble des éléments $(b_o,\ldots,b_n)$ de A^{n+1} vérifiant les relations (2) forment un sous-module de type fini de A^{n+1} ; par conséquent I est un idéal à gauche de type fini.

a) $\Rightarrow$ c) Soit P un élément de $A[X^n,\sigma]$, $E = E_{n-1}X^{n-1}+\ldots+E_o$ un idempotent qui engendre $\ell_{A[X,\sigma]}(P)$. E_oP est nul et Q appartient à $\ell_{A[X^n,\sigma]}(P)$ si et seulement si $Q = QE = QE_o$. E_o est donc un générateur de $\ell_{A[X^n,\sigma]}(P)$.

b) $\Rightarrow$ a) $A[X,\sigma]$ est un $A[X^n,\sigma]$-module à gauche libre de base $1,X,\ldots,X^{n-1}$. Soit $P = P_{n-1}X^{n-1}+\ldots+P_o$ un élément de $A[X,\sigma]$ avec $P_i \in A[X^n,\sigma]$ $(o\leq i\leq n-1)$. L'annulateur de cet élément est l'ensemble des polynomes de la forme $Q_{n-1}X^{n-1}+\ldots+Q_o$ tels que

$$(Q_{n-1}X^{n-1}+\ldots+Q_o)(P_{n-1}X^{n-1}+\ldots+P_o) = 0$$

c'est-à-dire tels que

$$\sum_{i+j\equiv k(n)} Q_jX^jP_iX^i = 0 \qquad (o\leq k\leq n-1)$$

L'ensemble des éléments $(Q_{n-1},\ldots,Q_o)$ vérifiant les relations précédentes est l'intersection des noyaux des applications linéaires f_k de $(A[X^n,\sigma])^n$ dans $A[X^n,\sigma]$ définies par

$$f_k((Q_{n-1},\ldots,Q_o)) = \sum_{i+j\equiv k(n)} Q_jX^jP_iX^i$$

$A[X^n,\sigma]$ étant semi-héréditaire à gauche, cette intersection est un sous-module de type fini de $(A[X^n,\sigma])^n$ engendré par les éléments $(Q_{n-1,i},Q_{n-2,i},\ldots,Q_{o,i})$ $(1\leq i\leq p)$. Par conséquent l'idéal $\ell_{A[X,\sigma]}(P)$ est engendré par les polynômes $Q_{n-1,i}X^{n-1}+\ldots+Q_{o,i}$ $(1\leq i\leq p)$ ∎

<u>COROLLAIRE</u>.- *Soient* A *un anneau régulier,* n *un entier,* σ *un automorphisme tel que* σ^n *soit intérieur. Alors les assertions suivantes sont équivalentes :*

a) $A[X,\sigma]$ *est semi-héréditaire à gauche.*

b) $A[X]$ *est semi-héréditaire à gauche.*

Il suffit de remarquer que si A est un anneau, σ_x l'automorphisme intérieur de A induit par l'élément inversible x ($\sigma_x(a) = x^{-1}ax$), alors l'application ϕ définie par

$$\phi : A[X,\sigma_x] \to A[X]$$
$$\Sigma a_i X^i \to \Sigma a_i x^{-i} X^i$$

est un isomorphisme d'anneaux ∎

THEOREME 1.- *Soit* A *un anneau régulier,* σ *un automorphisme de* A . *Alors* $A[X,\sigma]$ *est semi-héréditaire à gauche si, et seulement si, quels que soient les entiers* m *et* n *, l'annulateur à gauche dans* $M_{mn}(A)[X,\sigma^n]$ *d'un polynome de degré* 1 *est de type fini.*

COROLLAIRE.- *Soit* A *un anneau régulier.* $A[X]$ *est semi-héréditaire à gauche si et seulement si, pour tout* n *, l'annulateur dans* $M_n(A)[X]$ *d'un polynôme de degré* 1 *est un idéal à gauche de type fini.*

Avant de prouver ces résultats, on va donner une caractérisation commode des idéaux $C_n(I)$.

Soit $P = a_n X^n + \ldots + a_o$, $I = \ell_{A[X,\sigma]}(P)$ son annulateur. On considère les matrices n×n suivantes :

$$M = \begin{bmatrix} \sigma^{n-1}(a_n) & \sigma^{n-1}(a_{n-1}) \ldots\ldots \sigma^{n-1}(a_1) \\ & \sigma^{n-2}(a_n) \ \ldots\ldots \sigma^{n-2}(a_2) \\ & \ddots \quad \vdots \\ O & a_n \end{bmatrix} \qquad N = \begin{bmatrix} \sigma^{n-1}(a_o) & & O \\ \sigma^{n-2}(a_1) & \sigma^{n-2}(a_o) & \\ \vdots & \vdots \ \ddots & \\ a_{n-1} & a_{n-2} \ldots\ldots & a_o \end{bmatrix}$$

On prolonge σ à A^n en posant $\sigma[(x_1,\ldots,x_n)] = (\sigma(x_1),\ldots,\sigma(x_n))$ et on définit par récurrence une suite croissante (L_n) de sous-A-modules à gauche de A^n

$$L_o = (M:o) = \{Y \in A^n \ ; \ YM = o\}$$
$$L_p = (M:\sigma^n(L_{p-1}N)) = \{Y \in A^n \ ; \ YM \in \sigma^n(L_{p-1}N)\} .$$

<u>LEMME 5.</u>- $c_{np-1}(I)$ *est la* n-*ième projection de* $L_{p-1} \cap (N:0)$ *sur* A.

En effet b_o appartient à $c_{np-1}(I)$ si et seulement si il existe $b_{np-1},\dots,b_1$ tels que :

$$0 = (b_{np-1}X^{np-1}+\dots+b_o)(a_nX^n+\dots+a_o) \tag{3}$$

Pour tout entier $k \in [1,p]$ soit $Y_{p-k} = (b_{n(p-k+1)-1},\dots,b_{n(p-k)}) \in A^n$. Alors (3) équivaut à p relations dans A^n :

$$(4)\quad \begin{cases} 0 = Y_{p-1}\sigma^{n(p-1)}(M) \\ 0 = Y_{p-1}\sigma^{n(p-1)}(N) + Y_{p-2}\sigma^{n(p-2)}(M) \\ 0 = Y_{p-2}\sigma^{n(p-2)}(N) + Y_{p-3}\sigma^{n(p-3)}(M) \\ \dots\dots\dots\dots\dots\dots \\ 0 = Y_1\sigma^n(N) + Y_oM \\ 0 = Y_oN \end{cases}$$

En effet la (k+1)-ième, $0 = Y_{p-k}\sigma^{n(p-k)}(N) + Y_{p-k-1}\sigma^{n(p-k-1)}(M)$, s'obtient en identifiant dans (3) les coefficients de $X^{n(p-k+1)-1},\dots,X^{n(p-k)}$

Par définition Z appartient à L_{k-1} si et seulement si, il existe $Z' \in \sigma^n(L_{k-2})$ tel que $Z'\sigma^n(N) + ZM = 0$; donc $Y \in \sigma^{n(p-k)}(L_{k-1})$ équivaut à $Y'\sigma^{n(p-k+1)}(N) + Y\sigma^{n(p-k)}(M) = 0$ avec $Y' \in \sigma^{n(p-k+1)}(L_{k-2})$. Or la première relation de (4) signifie : $Y_{p-1} \in \sigma^{n(p-1)}(L_o)$ et la dernière : $Y_o \in (N:0)$; en conséquence $Y_o = (b_{n-1},\dots,b_o)$ appartient à $L_{p-1} \cap (N:0)$ si et seulement si, il existe $Y_{p-1},\dots,Y_1$ tels que $(Y_{p-1},\dots,Y_o)$ soit solution de (4) ∎

Maintenant, on peut démontrer le théorème 1. La proposition 1 montre que la condition est nécessaire. Réciproquement il suffit de prouver que si pour tout entier n , l'annulateur dans $M_n(A)[X,\sigma^n]$ d'un polynôme de degré 1 est de type fini, alors $A[X,\sigma]$ est un anneau de Rickart à gauche.

On reprend les notations du lemme 5 et on désigne par J l'annulateur du polynôme MX+N dans $M_n(A)[X,\sigma^n]$; par hypothèse J est un idéal de type fini et la suite $(C_p(J))$ est stationnaire.

Soit $B_o \in C_p(J)$. Il existe des éléments $B_p, \dots, B_1$ de $M_n(A)$ tels que les relations suivantes soient vérifiées

$$\begin{cases} 0 = B_p \sigma^{np}(M) \\ 0 = B_p \sigma^{np}(N) + B_{p-1} \sigma^{n(p-1)}(M) \\ \dots\dots\dots\dots\dots\dots \\ 0 = B_1 \sigma^{n}(N) + B_o M \\ 0 = B_o N \end{cases}$$

ou encore si $B_{k,i}$ désigne la i-ième ligne de la matrice B_k , pour tout i $(1 \leqslant i \leqslant n)$:

$$\begin{cases} 0 = B_{p,i} \sigma^{np}(M) \\ 0 = B_{p,i} \sigma^{np}(N) + B_{p-1,i} \sigma^{n(p-1)}(M) \\ \dots\dots\dots\dots\dots\dots \\ 0 = B_{1,i} \sigma^{n}(N) + B_{o,i} M \\ 0 = B_{o,i} N \end{cases}$$

On en déduit que B_o appartient à $C_p(J)$ si et seulement si, pour tout i , $B_{o,i}$ appartient à $L_p \cap (N:0)$ et par conséquent que la suite de A-modules $(L_p \cap (N:0))$ est stationnaire. D'après les lemmes 5) et 4) l'annulateur de P est un idéal de type fini ∎

Soit (A_α), $\alpha \in \Lambda$ une famille d'anneaux. Si :

$$\begin{array}{ccc} A_\alpha & \xrightarrow{u_{\beta\alpha}} & A_\beta \\ {\scriptstyle \sigma_\alpha}\downarrow & & \downarrow{\scriptstyle \sigma_\beta} \\ A_\alpha & \xrightarrow{u_{\beta\alpha}} & A_\beta \end{array}$$

est un système inductif filtrant où σ_α désigne un automorphisme de l'anneau A_α et $u_{\beta\alpha}$ sont des morphismes injectifs. On considère alors l'anneau $A = \varinjlim A_\alpha$ et l'automorphisme $\sigma = \varinjlim \sigma_\alpha$. On définit alors de façon évidente un système inductif $(A_\alpha[X,\sigma_\alpha], \tilde{u}_{\beta\alpha})$ dont la limite inductive est $A[X,\sigma]$.

LEMME 6.- *Soit* $(A_\alpha, u_{\beta\alpha})$ *un système inductif filtrant d'anneaux. On suppose* $u_{\beta\alpha}$ *injectif et* A_α *semi-héréditaire à gauche. Alors si* A_β *est un* A_α*-module à droite plat pour* $\beta>\alpha$, $A = \varinjlim A_\alpha$ *est un anneau semi-héréditaire à gauche.*

Pour démontrer ce lemme il suffit de reprendre l'exercice 12),e) chap.1, § 2 de [1] .

PROPOSITION 2.- *Soit* $(A_\alpha, u_{\beta\alpha}, \sigma_\alpha)$ *un système inductif filtrant d'anneaux réguliers. Si, pour tout* α , $A_\alpha[X,\sigma_\alpha]$ *est semi-héréditaire à gauche alors l'anneau* $A[X,\sigma] = \varinjlim A_\alpha[X,\sigma_\alpha]$ *est semi-héréditaire à gauche.*

Soit I un idéal à gauche de $A_\alpha[X,\sigma_\alpha]$. Soit $\beta>\alpha$; en posant

$$a_{i\beta}X^i \, . \, a_\beta \otimes x = a_{i\beta}\sigma_\beta^i\,(a_\beta) \otimes X^i x \, .$$

on munit $A_\beta \otimes_{A_\alpha} I$ d'une structure de $A_\beta[X,\sigma_\beta]$-module à gauche. Alors $a_{i\beta}X^i \otimes x \mapsto a_{i\beta} \otimes X^i x$ détermine un isomorphisme de $A_\beta[X,\sigma_\beta]$-modules à gauche entre $A_\beta[X,\sigma_\beta] \otimes_{A_\alpha[X,\sigma_\alpha]} I$ et $A_\beta \otimes_{A_\alpha} I$. On vérifie alors aisément que $A_\beta[X,\sigma_\beta]$ est un $A_\alpha[X,\sigma_\alpha]$-module à droite plat ∎

II.- DECOMPOSITION D'UN ANNEAU AUTO-INJECTIF REDUIT.

Dans ce paragraphe A désigne un anneau régulier réduit auto-injectif et σ un automorphisme de A . Pour tout idéal I de A on note $E(I)$ son enveloppe injective.

Définitions :

1) On appelle couverture σ-invariante de x et on note $C(x)$ l'idempotent (invariant par σ) qui engendre $E(\sum_{n\in\mathbb{Z}} A\,\sigma^n(x))$.

2) Deux idempotents e et f seront dits σ-isomorphes s'il existe $n\in\mathbb{Z}$ tel que $e = \sigma^n(f)$

3) A est <u>σ-fini</u> s'il vérifie les conditions équivalentes :

(i) tout idempotent e de A vérifie :

$$Ae \subset A\sigma(e) \Rightarrow e = \sigma(e)$$

(ii) un idempotent de A ne peut pas être orthogonal à tous ses transformés par σ .

4) A est <u>proprement σ-infini</u> s'il existe un idempotent e tel que la somme $\sum_{n \in \mathbb{N}} A\,\sigma^n(e)$ soit directe et essentielle dans A .

5) un idempotent e σ-invariant sera dit σ-fini (resp. proprement σ-infini) si l'anneau Ae est σ-fini (resp. proprement σ-infini).

<u>PROPOSITION 3</u>.- *Pour un anneau auto-injectif réduit, les assertions suivantes sont équivalentes :*

a) A *est* σ*-fini*

b) A *est* $<\sigma>$*-fini (i.e. pour tout* n *,* A *est* σ^n*-fini)*

a) $\Rightarrow$ b) il faut démontrer que A est σ^n-fini lorsque A est σ-fini, remarquons que pour $n = o$ l'assertion est évidente et d'autre part A est σ-fini si et seulement si A est σ^{-1}-fini car on a, si e est un idempotent de A

$$Ae \subset A\sigma(e) \iff A(1-e) \subset A\sigma^{-1}(1-e)$$

Il suffit de montrer la propriété pour n positif. On raisonne alors par récurrence sur n . Supposons démontré que pour tout automorphisme τ de A tel que A soit τ-fini on ait, pour tout entier p inférieur à n , A est τ^p-fini. Si n n'est pas premier, alors $n = pq$ avec $p<n$ et $q<n$. On a donc, si σ est un automorphisme de A , $\sigma^n = (\sigma^p)^q$ et dans ce cas, si A est σ-fini, il est σ^n-fini. On est donc amené au cas où n est premier. Supposons qu'il existe un automorphisme σ de A tel que A soit σ-fini mais ne soit pas σ^n-fini. Alors il existe un idempotent e de A tel que $Ae \subsetneq A\sigma^n(e)$ et on peut de plus supposer que Ae ne contient pas d'idempotents non nuls invariants par σ^n . En effet, soit

$\bigoplus_{i\in I} Ae_i$ une somme directe maximale d'idéaux σ^n-invariants, engendrés par des idempotents orthogonaux (e_i), contenue dans Ae ; l'enveloppe injective Af $(f=f^2)$ de la somme directe des Ae_i, $i\in I$ est invariante par σ^n et l'idempotent $g = e-f$ engendre un idéal ne contenant pas d'élément σ^n-invariant. De plus comme $f = \sigma^n(f)$ on a :

$$\sigma^n(e) = \sigma^n(g)+f$$

et $g = g\sigma^n(e) = g(f+\sigma^n(g)) = g\sigma^n(g)$ ce qui montre que $Ag \subsetneq A\sigma^n(g)$.

Si $Ag \subsetneq A\sigma^n(g)$ ne contient pas d'idempotents σ^n-invariants on a : $\prod_{j=o}^{n-1} \sigma^j(g) = 0$, sinon l'idempotent $\prod_{j=o}^{n-1} \sigma^j(g)$ serait invariant par σ (car A est σ-fini). De plus si ℓ désigne le plus petit entier tel que $\prod_{j=o}^{\ell} \sigma^j(g) = 0$ l'idempotent $\prod_{j=o}^{\ell-1} \sigma^j(g)$ est orthogonal avec au moins ses ℓ premiers transformés.

D'autre part si h est un idempotent tel que $Ah \subsetneq A\sigma^n(h)$ h ne peut être orthogonal avec ses $n-1$ premiers transformés car sinon l'idempotent $f = \sum_{j=o}^{n-1} \sigma^j(h)$ serait invariant par σ car A est σ-fini et on en déduirait que $h = \sigma^n(h)$.

Choisissons dans l'ensemble des idempotents e tels que $Ae \subsetneq A\sigma^n(e)$ et Ae ne contienne pas d'idempotent σ^n-invariant, un idempotent g tel que le nombre p des premiers transformés consécutifs orthogonaux à g soit maximum. On a donc

$$g\sigma^i(g) = o \qquad 1\leqslant i\leqslant p$$

et d'après les remarques précédentes $p<n-1$.

Nous allons construire un idempotent h tel que h ait les mêmes propriétés que g et soit de plus orthogonal à son $p+1$-ème transformé ce qui nous donnera une contradiction. n étant premier il existe deux entiers positifs α et β tels que :

$$\alpha(p+1) = \beta n+1$$

Soit $a = \prod_{j=o}^{\alpha} \sigma^{j(p+1)}(g)$, on a :

$$a = a\, g\, \sigma^{\alpha(p+1)}(g)$$

or $g = g\, \sigma^{n}(g) = g\sigma^{\beta n}(g)$

d'où $a = a\, g\, \sigma^{\beta n}(g)\, \sigma^{\alpha(p+1)}(g)$

$= a\, g\, \sigma^{\beta n}(g\sigma(g))$

Or par hypothèse $g\sigma(g) = 0$, d'où $a = 0$.

Soit ℓ le plus petit entier positif tel que

$$\prod_{j=o}^{\ell} \sigma^{j(p+1)}(g) = 0$$

L'idempotent $h = \prod_{j=o}^{\ell-1} \sigma^{j(p+1)}(g)$ est alors orthogonal à ses p+1 premiers transformés ce qui achève la démonstration ∎

THEOREME 2.- A *se décompose de manière unique en le produit d'un anneau* $\langle\sigma\rangle$*-fini et d'un anneau proprement* σ*-infini.*

Soit (e_i) une famille maximale d'idempotents proprement σ-infinis orthogonaux et soit e l'idempotent générateur de $E(\oplus Ae_i)$; il est proprement σ-infini. De plus $A(1-e)$ ne contient pas d'idempotent proprement σ-infini ; $A(1-e)$ est σ-fini. L'unicité est évidente ∎

Définition :

Un idempotent e de A est σ-abélien si Ae ne contient pas d'idempotents non nuls σ-isomorphes et orthogonaux.

LEMME 7.- *Soit* (e_i) *une famille d'idempotents* σ*-abéliens dont les couvertures* σ*-invariantes sont orthogonales. L'idempotent* e *générateur de* $E(\oplus Ae_i)$ *est* σ*-abélien.*

De la relation $C(e_j)e_i = \delta_{ij}\, e_j$ il résulte les égalités $AeC(e_j) = E(\oplus Ae_i C(e_j)) = Ae_j$ et $C(e_j)\, e = e_j$. Alors si f et $\sigma^n(f)$ sont des idempotents non nuls de Ae il existe j tel que $fC(e_j)$ ne soit pas nul et on a les égalités $fC(e_j) = fe_j$ et $\sigma^n(fC(e_j)) = \sigma^n(f)C(e_j) = \sigma^n(f)e_j$. e_j étant σ-abélien f et $\sigma^n(f)$ ne sont pas orthogonaux ∎

LEMME 8.- *Soient* e *et* f *deux idempotents. On suppose que* e *est* σ*-abélien et*

$$1 = \sum_{i=o}^{m-1} \sigma^i(e) = \sum_{j=o}^{n-1} \sigma^j(f)$$

Alors n *divise* m .

Soit $\alpha<n$ le plus petit entier tel que $e\sigma^\alpha(f)$ ne soit pas nul. Il existe deux entiers β et γ tels que $\alpha+m = n\beta+\gamma$ et $\gamma < n$. Des égalités $\sigma^m(e\sigma^\alpha(f)) = e\sigma^{m+\alpha}(f) = e\sigma^\gamma(f)$ il résulte que $e\sigma^\alpha(f)$ et $e\sigma^\gamma(f)$ sont σ-isomorphes. Comme e est σ-abélien, $\alpha = \gamma$ ∎

Définition :

1) On dit que A est de σ-type I s'il existe un idempotent σ-abélien e tel que $A = E(\sum_{n\in\mathbb{Z}} \sigma^n(e))$.

2) On dit que A est de σ-type I_n s'il existe un idempotent σ-abélien e orthogonal à ses $n-1$ premiers transformés tel que $1 = \sum_{i=o}^{n-1} \sigma^i(e)$.

LEMME 9.- A *est de* σ*-type* I *si et seulement si tout facteur direct* σ*-invariant contient un idempotent* σ*-abélien.*

La condition est suffisante car si (e_i) désigne une famille maximale d'idempotents σ-abéliens dont les couvertures sont orthogonales, d'après le lemme 7 l'idempotent e générateur de $E(\oplus Ae_i)$ est σ-abélien. De plus on a $C(e) = 1$ ∎

<u>LEMME 10</u>.- *Si* f *et* $\sigma^n(f)$ *sont des idempotents de* Ae *et si* e *est σ-abélien ils sont égaux.*

En effet les idempotents $f - f\sigma^n(f)$ et $\sigma^n(f) - \sigma^n(f)\sigma^{2n}(f)$ sont σ-isomorphes et orthogonaux donc nuls. De la même façon $\sigma^n(f) - \sigma^n(f)\, f$ est nul ∎

<u>THEOREME 3</u>.- *Tout anneau* A *de σ-type* I *σ-fini est isomorphe à un produit au plus dénombrable d'anneaux* A_{n_i} *de σ-type* I_{n_i} .

Il existe un idempotent e σ-abélien tel que $A = E(\sum_{n\in\mathbb{Z}} A\sigma^n(e))$ et un plus petit entier n tel que $e\,\sigma^n(e)$ ne soit pas nul. $\sigma^n(e)$ étant σ-abélien, le lemme 10 montre que $e\sigma^n(e) = \sigma^n(e\sigma^n(e))$ et $\sum_{i=o}^{n-1} A\sigma^i(e\sigma^n(e))$ est de σ-type I_n . Soit (e_α) une famille maximale d'idempotents de type I_n orthogonaux : si g_α désigne un idempotent σ-abélien tel que $e_\alpha = \sum_{i=o}^{n-1} \sigma^i(g_\alpha)$ et si g engendre $E(\oplus Ag_\alpha)$, g est σ-abélien (lemme 7) et $f_n = \sum_{i=o}^{n-1} \sigma^i(g)$ engendre $E(\oplus Ae_i)$ qui est donc le plus grand facteur direct de type I_n . De plus on a $A = E(\oplus_{n\in\mathbb{N}} Af_n)$ qui est isomorphe à $\prod_{n\in\mathbb{N}} Af_n$ ∎

<u>THEOREME 4</u>.- *Les assertions suivantes sont équivalentes :*

a) il existe N_1 *tel que pour tout idempotent* e *on ait* $e = \sigma^N(e)$

b) il existe N_2 *tel que les sommes directes d'idéaux engendrés par des idempotents σ-isomorphes soient de longueur inférieure à* N_2

c) A *se décompose en* $A_{n_1}\times\ldots\times A_{n_1}$ *où* A_{n_i} *est de σ-type* I_{n_i} .

b) ⟹ c) Soit k le plus grand entier tel qu'il existe un idempotent e orthogonal à ses k premiers transformés $\sigma(e),\ldots,\sigma^k(e)$. $e - e\sigma^{k+1}(e)$ et $\sigma^{k+1}(e) - \sigma^{k+1}(e)e$ étant orthogonaux à leurs k+1 premiers transformés,

e est égal à $\sigma^{k+1}(e)$. De plus tout idempotent $f \in Ae$ est orthogonal à ses k premiers transformés donc $f = \sigma^{k+1}(f)$ et e est σ-abélien. Tout facteur direct invariant de A vérifie b) ; d'après le lemme 9 A est de σ-type I σ-fini. c) résulte alors du théorème 3.

c) $\Rightarrow$ a) Si A est de (σ)-type I_n et si $1 = \sum_{i=o}^{n-1} \sigma^i(f)$ avec f σ-abélien, pour tout idempotent e l'égalité $\sigma^n(\sigma^i(f)e) = \sigma^i(f)\sigma^n(e)$ et le lemme 10 entrainent $\sigma^n(\sigma^i(f)e) = \sigma^i(f)e$ et $e = \sigma^n(e)$ ∎

Définition :

Un anneau qui vérifie les conditions du théorème 4 sera dit de (σ)-type I borné.

III.- APPLICATIONS DE LA DECOMPOSITION.

Soient A un anneau régulier, $\mathbb{Z}$ le groupe des entiers relatifs. On définit sur $A^{\mathbb{Z}}$ un automorphisme d en posant :

$$d\left[(a_n)_{n\in\mathbb{Z}}\right] = (a_{n-1})_{n\in\mathbb{Z}} .$$

THEOREME 5.- *Soient* A *un anneau régulier,* d *l'automorphisme de* $A^{\mathbb{Z}}$ *défini ci-dessus. L'anneau* $A^{\mathbb{Z}}[X,d]$ *n'est pas semi-héréditaire à gauche.*

Soit $a = (a_k)_{k\in\mathbb{Z}}$, un élément de $A^{\mathbb{Z}}$. La relation : $(c_pX^p+\ldots+c_o)\ (Xa+a) = o$ équivaut à :

$$\begin{cases} o = c_p d^{p+1}(a) \\ o = (c_p+c_{p-1})d^p(a) \\ \cdot\ \cdot\ \cdot\ \cdot\ \cdot\ \cdot\ \cdot\ \cdot\ \cdot\ \cdot \\ o = (c_{p-k}+c_{p-k-1})d^{p-k}(a) \\ \cdot\ \cdot\ \cdot\ \cdot\ \cdot\ \cdot\ \cdot\ \cdot\ \cdot\ \cdot \\ o = (c_1+c_o)d(a) \\ o = c_o a \end{cases}$$

ou encore à :

$$\forall n \in \mathbb{Z} \quad \begin{cases} o = c_{p,n}a_{n-(p+1)} \\ o = (c_{p,n}+c_{p-1,n})a_{n-p} \\ \cdots\cdots\cdots \\ o = (c_{p-k,n}+c_{p-k-1,n})a_{n-(p-k)} \\ \cdots\cdots\cdots \\ o = (c_{1,n}+c_{o,n})a_{n-1} \\ o = c_{o,n}a_n \end{cases} \qquad (1)$$

Si $x = (x_k)_{k\in\mathbb{Z}}$ appartient à $A^{\mathbb{Z}}$ on définit le support de x par : $\mathrm{Supp}\, x = \{k/k\in\mathbb{Z}\ ,\ x_k \neq o\}$. Soit $a = (a_k)_{k\in\mathbb{Z}}$ l'élément de $A^{\mathbb{Z}}$ défini par :

$a_k=1$ s'il existe un entier $n\geq o$ tel que $\frac{(n+2)(n+3)}{2} - 2\leq k\leq \frac{(n+2)(n+3)}{2}+n-2$

$a_k = 0$ sinon.

Considérons l'élément c de $A^{\mathbb{Z}}$ défini par :

$c_k = 1$ si $k\notin \mathrm{Supp}\, a$, $k-i\in \mathrm{Supp}\, a$ $(1\leq i\leq p)$, $k-(p+1)\notin \mathrm{Supp}\, a$

$c_k = o$ sinon.

Alors on a : $c(\sum_{i=o}^{p} (-1)^i X^i)\ (Xa+a) = 0$, ce qui veut dire que $c\in C_p(\ell(Xa+a))$. Les relations (1) montrent d'autre part que si $p'<p$, c n'appartient pas à $C_{p'}(\ell(Xa+a))$. La suite $C_p(\ell(Xa+a))$, $p\in\mathbb{N}$ est strictement croissante ∎

<u>LEMME 11</u>.- *Soit* $(A_n), n\in\mathbb{Z}$, *une famille d'anneaux. Si* σ *est un automorphisme de l'anneau* $\prod_{n\in\mathbb{Z}} A_n$ *tel que pour tout entier* p $\sigma(A_p) = A_{p+1}$, *alors l'anneau* $\prod_{n\in\mathbb{Z}} A_n[X,\sigma]$ *est isomorphe à l'anneau* $A_o^{\mathbb{Z}}[X,d^{-1}]$.

Comme $A_n = \sigma^n(A_o)$, tout élément x de l'anneau $\prod_{n\in\mathbb{Z}} A_n$ se met sous la forme $x = (\sigma^n(a_n))_{n\in\mathbb{Z}}$ avec $a_n\in A_o$ pour tout entier n .

Soit $P = \Sigma x_p X^p$ un élément de $\prod_{n\in\mathbb{Z}} A_n[X,\sigma]$ avec $x_p = (\sigma^n(a_{n,p}))_{n\in\mathbb{Z}}$.
L'application ϕ de $\prod_{n\in\mathbb{Z}} A_n[X,\sigma]$ dans $A_o^{\mathbb{Z}}[X,d^{-1}]$ définie par

$$\phi(P) = \Sigma(a_{n,p})_{n\in\mathbb{Z}} X^p$$

est un isomorphisme d'anneaux ∎

Des résultats précédents on déduit le :

COROLLAIRE.- *Soient* A *un anneau régulier réduit auto-injectif,* σ *un automorphisme de* A . *Si* $A[X,\sigma]$ *est semi-héréditaire à gauche,* A *est* σ*-fini.*

Si A n'est pas σ-fini, il admet un facteur direct σ-invariant, proprement infini. On peut donc supposer que A est proprement infini : il existe un idempotent e tel que $A = \prod_{n\in\mathbb{Z}} A\sigma^n(e)$. Le lemme 11 et le théorème 5 permettent alors de conclure ∎

LEMME 12.- *Soient* A *un anneau,* Int x *un automorphisme intérieur de* A , σ *un automorphisme de* A . *L'anneau* $A[X,\text{Int } x \circ \sigma]$ *est isomorphe à l'anneau* $A[X,\sigma]$.

L'application ϕ de $A[X,\text{Int } x \circ \sigma]$ dans $A[Y,\sigma]$ définie par $\phi(\sum_{i=o}^{n} a_i X^i) = \sum_{i=o}^{n} \alpha_i Y^i$ où α_i est déterminé par $a_i X^i = \alpha_i (xX)^i$ est un isomorphisme d'anneaux. C'est évidemment un homomorphisme de groupes et comme

$$\phi(aX^i b) = ax^{-1}\sigma(x^{-1})\ldots\sigma^{i-1}(x^{-1})\sigma^i(b)Y^i = \phi(aX^i)\phi(b)$$

c'est un homomorphisme d'anneaux ∎

La proposition suivante est due à G. Renault.

PROPOSITION 4.- *Soient* A *un anneau régulier auto-injectif réduit,* n *un entier. Tout automorphisme de* $M_n(A)$ *est produit d'un automorphisme intérieur par un automorphisme de* A .

Soit σ un automorphisme de $M_n(A) = \text{End}_A(A^n)$, et soit $\{e_1,\dots,e_n\}$ la base canonique de A^n , (p_i), $1\leq i\leq n$ la famille des projecteurs associée à cette base, E_{ij} les n^2-éléments de $M_n(A)$ définis par $E_{ij}(e_j) = e_i$, $E_{ij}(e_k) = o$ pour $k \neq j$.

Alors la famille $\sigma(p_i)$, $1\leq i\leq n$ est un ensemble de n projecteurs orthogonaux deux à deux dont la somme est 1 .

Comme A est fini on a

$$A^n = \bigoplus_{i=1}^{n} \text{Im } \sigma(p_i)$$

avec $\text{Im } \sigma(p_i) \sim A$. Posons $\text{Im } \sigma(p_i) = Af_i$. D'autre part les éléments $U_{ij} = \sigma(E_{ij})$ forment un système d'unités de $M_n(A)$ associé à la base $\{f_i \ , 1\leq i\leq n\}$. L'endomorphisme x de A^n défini par

$$x(f_i) = e_i \qquad 1\leq i\leq n$$

est un isomorphisme tel que

$$U_{ij} = x^{-1}E_{ij}\,x$$

Par conséquent l'automorphisme $\text{Int } x \circ \sigma$ de $M_n(A)$ laisse les matrices E_{ij} invariantes, cet automorphisme provient donc d'un automorphisme de A ∎

<u>COROLLAIRE</u>.- *Soit* $A = \prod_{n\in\mathbb{N}} M_n(D_n)$ *un anneau régulier auto-injectif de type* I *fini, où, pour tout entier* n , D_n *est un anneau régulier réduit auto-injectif* [6] . *Si* σ *est un automorphisme de* A , *il existe des automorphismes* σ_n *de* D_n *tels que* σ *soit produit d'un automorphisme intérieur de* A *par l'automorphisme* $\prod_{n\in\mathbb{N}} \sigma_n$.

Montrons que $Ah_n = M_n(D_n)$ est invariant par σ . Si $h_n = \sum_{i=1}^{n} e_i$ où les e_i sont des idempotents abéliens orthogonaux deux à deux isomorphes on a

$$\sigma(h_n) = \sum_{i=1}^{n} \sigma(e_i)$$

où les $\sigma(e_i)$ sont des idempotents abéliens orthogonaux deux à deux isomorphes, comme Ah_n est le composant homogène de type I_n de A on a : $Ah_n = A\sigma(h_n)$, et σ induit un automorphisme sur chaque $M_n(D_n)$, la proposition précédente donne alors le résultat.

PROPOSITION 4.- *Soient* A *un anneau régulier auto-injectif de type* I *fini,* σ *un automorphisme de* A *. Si* $A[X,\sigma]$ *est semi-héréditaire à gauche,* A *est de type* I *borné (i.e.* $A[X]$ *est semi-héréditaire à gauche).*

D'après le lemme 12 et le corollaire de la proposition 3 , on peut supposer que si $A = \prod_{n\in\mathbb{N}} M_n(D_n)$, $\sigma = \prod\sigma_n$ où σ_n est un automorphisme de D_n . Une adaptation facile du lemme 10 de [3] et du théorème 12 de [3] , donne alors le résultat ∎

LEMME 13.- *Soient* $A = \prod_{i=o}^{n} A_i$ *un anneau,* σ *un automorphisme de* A *tel que* $\sigma(A_i) = A_{i+1}$ $o \leq i \leq n-1$, $\sigma(A_n) = A_o$ *. Alors dans l'anneau* $A[X,\sigma]$ *les idéaux* $\ell((0,1,1,\dots,1)X + (1,1,\dots,1,0))$ *et* $r(X(1,1,\dots,1,0) + (0,1,1,\dots,1))$ *ne sont pas nuls et ne contiennent pas de polynôme de degré strictement inférieur à* n .

Il suffit de reprendre la démonstration du théorème 5 avec $a = (1,1,\dots,1,0)$ et $c = (0,\dots,0,1)$ les indices étant pris modulo $n+2$ ∎

THEOREME 6.- *Soient* A *un anneau régulier réduit auto-injectif,* σ *un automorphisme de* A *. Si* A *est de* σ*-type* I σ*-fini,* $A[X,\sigma]$ *est semi-héréditaire à gauche si et seulement si* A *est de* σ*-type* I *borné.*

Si A est de σ-type I σ-fini, alors d'après le théorème 3 $A = \prod_{n \in \mathbb{N}} A_n$ où A_n est de σ-type I_n , il suffit alors de reprendre la démonstration du théorème 12 de (3) en remplaçant le lemme 10 de (3) par le lemme 13 pour obtenir le résultat. Réciproquement si A est de σ-type I σ-borné, d'après le théorème 4 , il existe un entier N tel que σ^N laisse les idempotents de A invariants ; d'après la proposition 1, il suffit de prouver le lemme suivant :

LEMME 14.- *Soient* A *un anneau réduit,* σ *un automorphisme de* A . *Si les idempotents de* A *sont* σ*-invariants,* $A[X,\sigma]$ *est semi-héréditaire à gauche (et à droite). De plus* $A[X,\sigma]$ *est un anneau de Bezout.*

La démonstration est identique à celle de la proposition 6 de (3) ∎

IV.- QUELQUES EXEMPLES D'ANNEAUX SEMI-HEREDITAIRES.

Soient A_i , $i \in I$ une famille d'anneaux simples et σ un automorphisme de l'anneau $A = \prod_{i \in I} A_i$. Les images par σ des A_i sont des idéaux bilatères minimaux de A . Nous dirons que deux indices i et j de I sont équivalents s'il existe un entier relatif k tel que $\sigma^k(A_i) = A_j$, on notera c(i) le cardinal de l'ensemble des éléments équivalents à i . On notera $\ell(i)$ la longueur du socle de A_i .

THEOREME.- *Soient* A_i , $i \in I$ *une famille d'anneaux simples,* σ *un automorphisme de* $A = \prod_{i \in I} A_i$. $A[X,\sigma]$ *est semi-héréditaire à gauche si et seulement si il existe un entier* N *tel que*

1) $c(i) \leq N$ *pour tout* $i \in I$.

2) $\ell(i) \leq N$ *pour tout* $i \in I$.

Supposons $A[X,\sigma]$ semi-héréditaire à gauche, la proposition 4 montre qu'il existe un entier N_1 tel que $\ell(i) \leq N_1$ pour tout élément i de I, et le théorème 6 l'existence d'un entier N_2 tel que $c(i) \leq N_2$ pour tout élément i de I. Réciproquement d'après la condition 2) $A = \prod_{i=1}^{N} M_i(D_i)$ où les anneaux D_i sont des produits de corps et $\sigma = \prod_{i=1}^{n} \sigma_i$ où σ_i est un automorphisme de $M_i(D_i)$. On est donc ramené au cas où $A = \prod_{j\in I} M_n(K_j)$, où les K_j sont des corps. La condition 1) montre alors qu'il existe un entier p tel que σ^p soit produit d'automorphismes τ_j de $M_n(K_j)$, il suffit alors, d'après la proposition 1, de considérer le cas où $A = \prod_{j\in J} M_n(K_j)$ et $\sigma = \Pi\sigma_j$ où σ_j est un automorphisme de $M_n(K_j)$. D'après la proposition 3 chaque automorphisme σ_j est produit d'un automorphisme intérieur par un automorphisme τ_j de K_j, le lemme 12 montre qu'alors $A[X,\sigma] \sim A[X,\prod_j \tau_j]$. Or on a :

$$A[X,\Pi\tau_j] \simeq M_n\left[(\prod_j K_j)[X,\Pi\tau_j]\right]$$

et le lemme 14 permet de conclure ∎

Nous avons vu que si σ est un automorphisme d'un anneau régulier réduit A tel que $\sigma^n = 1$, pour un entier n, alors l'anneau $A[X,\sigma]$ est semi-héréditaire à gauche, l'exemple suivant montre qu'on ne peut espérer de condition d'algébricité pour l'automorphisme.

Soit A un anneau semi-simple, $\mathcal{S}(A)$ le sous-anneau de $A^{\mathbb{Z}}$ formé des suites constantes à partir d'un certain rang c'est-à-dire :

$$\mathcal{S}(A) = \{(a_n)_{n\in\mathbb{Z}}\ ,\ \exists\, r\in\mathbb{N}\quad \forall \ell\in\mathbb{Z}\ ,\ |\ell| \geq r \Rightarrow a_\ell = a_r\}.$$

Soit σ l'automorphisme de $\mathcal{S}(A)$ défini par :

$$\sigma((a_n)_{n\in\mathbb{Z}}) = (b_n)_{n\in\mathbb{Z}} \quad \text{avec } b_n = a_{n-1}$$

On vérifie facilement que l'application f de $\mathcal{S}(A)$ dans $\mathcal{S}(A^p)$ définie par $f((a_n)_{n\in\mathbb{Z}}) = (a_{np+1},\dots,a_{(n+1)p})_{n\in\mathbb{Z}}$ est un isomorphisme d'anneaux et que :

$$f(\sigma^p(a)) = \sigma(f(a))$$

ce qui montre que l'application ϕ de $\mathcal{S}(A)[X,\sigma^p]$ dans $\mathcal{S}(A^p)[X,\sigma]$ définie par :

$$\phi(\sum_{i=1}^{n} \alpha_i X^i) = \sum_{i=1}^{n} f(\alpha_i)X^i$$

est un isomorphisme d'anneaux.

<u>PROPOSITION 5.-</u> *Soit* A *un anneau semi-simple. L'anneau* $\mathcal{S}(A)[X,\sigma]$ *est semi-héréditaire à gauche.*

En vertu du théorème 1 il suffit, d'après les remarques précédentes, de montrer que l'annulateur à gauche dans $\mathcal{S}(A)[X,\sigma]$ d'un polynôme de degré 1 est de type fini.

Soit MX+N un élément de $\mathcal{S}(A)[X,\sigma]$ avec $M = (M_p)$, $p\in\mathbb{Z}$ et $N = (N_p)$, $p\in\mathbb{Z}$. On peut supposer que pour $p \leqslant 0$ et $p \geqslant r+1$ (r fixé) $M_p = m$ et $N_p = n$. Si $I = \ell(MX+N)$ il suffit de démontrer que la suite $C_p(I)$, $p\in\mathbb{N}$ est stationnaire, c'est-à-dire que la suite $L_p(I) = L_p$ est stationnaire (car $C_p(I) = L_p(I) \cap (N:0)$) . Soit $L_{p,i}$ la i-ème projection de L_p sur A alors pour tout $i < o$ et pour tout $i \geqslant r+1$ on a :

$$L_{p+1,i+1} = (m:L_{p,i}n)$$

et les diagrammes suivants (les flèches désignent des inclusions)

$$\begin{array}{ccccccc}
\cdots = L_{o,-2} & = & L_{o,-1} & = & L_{o,o} & = & L_{o,r+1} \\
\downarrow & & \downarrow & & \downarrow & & \\
\cdots = L_{1,-2} & = & L_{1,-1} & = & L_{1,o} & = & L_{1,r+2} \\
\downarrow & & \downarrow & & \downarrow & & \\
\cdots = L_{2,-2} & = & L_{2,-1} & = & L_{2,o} & = & L_{2,r+3} \\
\downarrow & & \downarrow & & \downarrow & & \\
\vdots & & \vdots & & \vdots & & \\
\downarrow & & \downarrow & & \downarrow & & \\
\cdots = L_{s-1,-2} & = & L_{s-1,-1} & = & L_{s-1,o} & = & L_{s-1,r+s} \\
\downarrow & & \downarrow & & \downarrow & &
\end{array}$$

et

$$\begin{array}{cccccccccc}
L_{o,r+1} & = & L_{o,r+2} & = & L_{o,r+3} & = \cdots & & \cdots = & L_{o,r+s} & = \cdots \\
\downarrow & \searrow & \downarrow & & \downarrow & & & & \downarrow & \\
L_{1,r+1} & & L_{1,r+2} & = & L_{1,r+3} & = \cdots & & \cdots = & L_{1,r+s} & = \cdots \\
\downarrow & & \downarrow & \searrow & \downarrow & & & & \downarrow & \\
L_{2,r+1} & & L_{2,r+2} & & L_{2,r+3} & = \cdots & & \cdots = & L_{2,r+s} & = \cdots \\
\downarrow & & \downarrow & & \downarrow & \searrow & & & \downarrow & \\
\vdots & & \vdots & & \vdots & & \ddots & & \vdots & \\
\downarrow & & \downarrow & & \downarrow & & & \searrow & \downarrow & \\
L_{s-1,r+1} & & L_{s-1,r+2} & & L_{s-1,r+3} & & & & L_{s-1,r+s} & = \cdots \\
\downarrow & & \downarrow & & \downarrow & & & & \downarrow & \searrow
\end{array}$$

Montrons que si A est de longueur k , alors quel que soit p et quel que soit $s \geqslant k$ on a :

$$(n:o) \cap L_{p+s,r+s+1} = (n:o) \cap L_{k,r+k+1}$$

On a les inclusions suivantes :

$$\begin{array}{ccccccccc}
I_o = L_{o,r+1} & \to & I_1 = L_{1,r+2} & \to & I_2 = L_{2,r+3} & \to\dots\to & I_{n-1} = L_{n-1,2+n} & \to & I_n = L_{n,r+n+1} \\
\downarrow & & \downarrow & & \downarrow & & \downarrow & & \downarrow \\
L_{p,r+1} & & L_{p+1,r+2} & & L_{p+2,r+3} & & L_{p+n-1,r+n} & & L_{p+n,r+n+1} \\
\downarrow & & \downarrow & & \downarrow & & \downarrow & & \downarrow \\
J_o = A & \leftarrow & J_1=(m:J_on) & \leftarrow & J_2=(m:J_1n) & \leftarrow\dots\leftarrow & J_{n-1} = (m:J_{n-2}n) & \leftarrow & J_n = (m:J_{n-1}n)
\end{array}$$

Il suffit donc de montrer que $(n:o) \cap I_k = (n:o) \cap J_k$. On définit une suite croissante d'idéaux à gauche de A par :

$$K_\ell = \{x \in A \ , \ \exists(x_\ell, x_{\ell-1}, \dots, x_1, x_o) \quad x_k n = o \ , \ x_k m = x_{k-1} n \dots, x_1 m = xn\}$$

Il existe un entier $\ell \leqslant k$ tel que $K_{\ell-1} = K_\ell$. Or x appartient à $J_\ell \cap (n:o)$ si et seulement si il existe des éléments $y_o, \dots, y_{\ell-1}$ tels que

$$xn = o \ , \ xm = xy_{\ell-1}, \dots \qquad , \ y_1 m = y_o n \qquad (1)$$

De sorte que y_o appartient à $K_\ell = K_{\ell-1}$, il existe donc des éléments $y'_{\ell-1}, \dots, y'_o = y_o$ tels que

$$o = y'_{\ell-1} n \ , \ y'_{\ell-1} m = y'_{\ell-2} n, \dots, y'_1 m = y'_o n \qquad (2)$$

D'où en retranchant (2) de (1) on obtient les égalités :

$$xn = o \qquad xm = (y_\ell - y'_{\ell-1})n, \dots, (y_1 - y'_1)m = o$$

ce qui montre que x appartient à $(n:o) \cap I_{k-1}$ ∎

<u>DEFINITION</u>.- *Soit* A *un anneau,* a *et* b *deux éléments de* A *. On définit par récurrence une suite* $(L_n(a,b))$, $n \in \mathbb{N}$ *en posant :*

$$L_o(a,b) = (a:o)$$

$$L_n(a,b) = (a:L_{n-1}(a,b)b) = \{x / xa \in L_{n-1}(a,b)b\} \quad \text{pour} \quad n \geqslant 1$$

La suite ainsi définie est une suite croissante d'idéaux à gauche de A .

<u>PROPOSITION 6</u>.- *Si* A *est un* (Σ-V)*-anneau régulier*, $A[X]$ *est semi-héréditaire à gauche (et à droite).*

Soient a et b deux éléments de $M_p(A)$ où p désigne un entier, nous noterons pour simplifier $L_n = L_n(a,b)$. L'application ϕ_o de L_1 dans $L_o b$ définie par $\phi_o(x) = xa$ a pour noyau L_o , il existe donc un homomorphisme injectif de L_1/L_o dans L_o . Montrons que pour tout entier n il existe un homomorphisme injectif de L_n/L_{n-1} dans L_{n-1}/L_{n-2} . Soit ϕ_n l'application de L_n/L_{n-1} dans $L_{n-1}b/L_{n-2}b$ définie par $\phi_n(\overline{x}) = \overline{xa}$, ϕ_n est bien une application car si $\overline{x} = \overline{y}$ alors $x-y \in L_{n-1}$ et $(x-y)a \in L_{n-2}b$, c'est donc un homomorphisme de manière évidente. D'autre part on a :

$$L_{n-1} = L_{n-2} \oplus L'_{n-1} \qquad (1)$$

d'où $$L_{n-1}b = L_{n-2}b + L'_{n-1}b$$

et par conséquent $$L'_{n-1}b \gtrsim L_{n-1}b/L_{n-2}b$$

d'où $$L_{n-1}b/L_{n-2} \lesssim L_{n-1}/L_{n-2}$$

Il en résulte que la suite (L'_{n-1}) , $n \in \mathbb{N}$ définit une somme directe d'idéaux ayant la propriété :

$$\forall\, n \in \mathbb{N} \qquad L'_n \gtrsim L'_{n-1}$$

donc si A est un (Σ-V)-anneau, il en est de même de $M_p(A)$ et la suite (L_n) , $n \in \mathbb{N}$ est stationnaire, ce qui montre à l'aide du théorème 1 que $A[X]$ est semi-héréditaire à gauche ∎

BIBLIOGRAPHIE

[1] N. BOURBAKI - Algèbre commutative, chap. 1 & 2. Hermann
[2] J. B. CASTILLON - C.R.A.S Paris 269 (1969) 556-559.
[3] J.M. GOURSAUD & J.L. PASCAUD - Anneaux semi-héréditaires. C.R. Acad. Sc. Paris, t. 284, série A, 1977, p 583.
[4] H. LENZING - Math. Zeit, 118, 1970, p.219-240.
[5] P.J. MAC CARTHY - Proc. Amer Math. Soc 39 (1973) 253-254.
[6] G. RENAULT - Bull Soc. Math. 101, 1973, p. 237-254.
[7] J.P. SOUBLIN - J. of Algebra, 15, 1970, p. 455-472.

LOWER K-THEORY, REGULAR RINGS AND OPERATOR ALGEBRAS - A SURVEY

David Handelman[(1)]

University of Ottawa, Ottawa, Canada

John Lawrence[(2)]

University of Waterloo, Waterloo, Canada.

§1. Introduction.

Regular rings were invented by von Neumann in 1936 in a brief paper in the Proceedings of the National Academy of Sciences. His motivation for the definition and the study of these rings was the (then) recent work that he had done jointly with F. J. Murray on rings of operators (now known as von Neumann algebras or W*-algebras) on a Hilbert space. They showed that the lattice of projections of finite von Neumann algebras formed a continuous geometry and von Neumann had showed that such continuous geometries could be coordinatized by a unique regular ring. He apparently hoped that the study of the regular ring associated with a von Neumann algebra would enhance our knowledge of the algebra.

For the most part, regular rings have failed in their original purpose. Although there are a great number of people now studying operator algebras, only a handful are interested in, or aware of, the algebraic methods and algebraic problems in this area.

In this article, we have tried to survey the applications of regular rings to the study of operator algebras. We have put particular

(1) Research partially supported by NRC Grant A3130.

(2) Research partially supported by NRC Grant A4540 and University of Waterloo Research Grant 131-7052.

emphasis on fairly recent work, in the hope that this will motivate others to explore this area.

For more information on regular rings, we direct the reader to K. Goodearl's forthcoming book 'Von Neumann Regular Rings'.

§2. Regular Rings.

A (von Neumann) regular ring is an associative ring with unity with the property that for every x there is a y such that $xyx = x$. In this case the element $e = xy$ is an idempotent and x and e generate the same right ideal in the ring.

Usually, several other conditions must be added to a regular ring before we can say very much about it. Two important classes of regular rings are directly finite regular rings and unit regular rings. A ring is directly finite if $ab = 1$ implies $ba = 1$ for all elements a and b in the ring. A ring is unit regular [7] if for every x there exists a unit u such that $xux = x$. In order to discuss the significance of unit regular rings we introduce K_0 (the Grothendieck group) of a regular ring.

Let R be a ring and let $\mathbb{P}$ be the set of isomorphism classes of finitely generated projective right R-modules. We construct the abelian group $K_0(R)$ with generators $\{[P], P\in\mathbb{P}\}$ and relations $[P]=[Q]+[S]$ for every short exact sequence $0\to Q\to P\to S\to 0$ of projectives. In general $[P] = [Q]$ in $K_0(R)$ does not imply that $P\cong Q$ as modules; however, if R is unit regular, the above implication is true.

Theorem . [13] Let R be a regular ring. The following properties are equivalent

a) R is unit regular

b) If A, B and C are finitely generated projective modules and $A\oplus B \cong A\oplus C$, then $A \cong C$.

c) If $[A] = [B]$ in $K_0(R)$, then $A \cong B$.

The importance of K_0 in the study of regular rings arises from the fact that under reasonably mild assumptions, it is a partially ordered abelian group. Suppose that R is regular and that it and all its matrix rings are directly finite. We define a partial order on $K_0(R)$ by letting

$$[A]-[B] \geq [C]-[D]$$

if there exists a finitely generated projective P such that

$$B \oplus C \oplus P \lesssim A \oplus D \oplus P$$

If P is a finitely generated projective, then

$$P \lesssim \underbrace{R \oplus R \oplus \ldots \oplus R}_{n\text{-times}}$$

for some natural number n. An element μ in a directed abelian group $(G,\geq)$ is an <u>order</u> <u>unit</u> if for every $x\in G$, there exists a natural number n such that $n\mu\geq x$. Thus $K_0(R)$ is a directed abelian group with order unit $[R]$. We denote the pair by $(K_0(R),[R])$.

We next consider functionals

$$f: (K_0(R),[R]) \to \mathbb{R};$$

these are order preserving group homomorphisms sending $[R]$ to 1 ($\mathbb{R}$ is the reals). If R is regular and all its matrix rings are directly finite, such a functional exists, the proof being similar to that of the Hahn-Banach Theorem. Given such a functional f we define a function $N: R \to \mathbb{R}$ by $N(a) = f([aR])$.

<u>Theorem</u> Suppose R is regular and all its matrix rings are directly finite. Then there exists a function $N: R \to \mathbb{R}$ satisfying

a) $N(1) = 1$

b) $N(xy) \le N(x), N(y)$, for all $x, y \in R$,

c) $N(e+f) = N(e)+N(f)$, for all orthogonal idempotents $e, f \in R$

Such a function is called a <u>pseudo-rank function</u>. If a pseudo-rank function also satisfies

$$N(x) = 0 \text{ if and only if } x = 0,$$

then it is called a <u>rank function</u>. A corollary of the above theorem is that all simple unit regular rings have rank functions.

§3. <u>Operator Algebras</u>.

Let H be a Hilbert space with the bounded (continuous) operators on H denoted by $\mathbb{B}(H)$. There is an involution $*$ on $\mathbb{B}(H)$ given by the adjoint function. If $\{S_n\}$ is a net of operators and S

is another operator then

a) $\{S_n\}\to S$ in the uniform topology if $||S_n-S||\to 0$ as $n\uparrow$.

b) $\{S_n\}\to S$ in the strong topology if $|S_nx-Sx|\to 0$, for all $x\in H$, as $n\uparrow$,

c) $\{S_n\}\to S$ in the weak topology if $(S_nx-Sx,y)\to 0$ for all $y,x\in H$, as $n\uparrow$.

A subalgebra $A\subset B(H)$ which is closed under $*$ and the weak operator topology is called a <u>von Neumann algebra</u> (also called a W*-algebra). If $S\subset B(H)$, let $S' = \{x\in B(H): xs=sx$ for all $s\in S\}$. Then a subalgebra $A\subset B(H)$ closed under $*$, which satisfies the double commutant condition $A'' = A$ is a von Neumann algebra, and in fact, this is simply another characterization of von Neumann algebras.

More general than the von Neumann algebras are the <u>C*-algebras</u>. These are self-adjoint subalgebras of $B(H)$ closed under the uniform topology. Gelfand found an internal characterization of C*-algebras which, in its final form, states that a C*-algebra is a Banach algebra in which

$$||x^*x|| = ||x^*||\ ||x||$$

for all x.

In addition to being a C*-algebra, a von Neumann algebra has the algebraic property that the right annihilator of every set is generated (as a right ideal) by a projection (i.e. a self-adjoint idempotent).

In [20], Kaplansky called a C*-algebra with such an annihilator property, an <u>AW*-algebra</u> (abstract W*-algebra). He showed in this paper that most of the algebraic properties that von Neumann and Murray had established for von Neumann algebras also hold for the more general AW*-algebras.

Suppose e and f are two projections in an AW*-algebra A. We say that e is <u>equivalent</u> to f, $e \sim f$, if they generate isomorphic right ideals, or equivalently, if there exist x and y such that $e=xy$ and $f = yx$. We say $e \leq f$ if $e = ef$ (= fe). A projection $e \in A$ is <u>abelian</u> if all idempotents in eAe are central. A projection $e \in A$ is <u>finite</u> if eAe is directly finite ($xy = 1$ implies $yx = 1$).

<u>Theorem</u> [21]. If A is an AW*-algebra and $x \in A$, then the right projection of x is equivalent to the left projection of x.

Following Murray and von Neumann, Kaplansky divided the AW*-algebras into five types. He proved that any AW*-algebra can be uniquely decomposed into a direct product of the five types.

An AW*-algebra A is <u>type I</u> if it has a faithful abelian idempotent. A is of <u>type II</u> if it has a faithful finite idempotent but no nonzero abelian idempotents A is of <u>type III</u> if it has no nonzero finite idempotents. Finally, A is <u>purely infinite</u> if it has no nonzero central finite idempotents.

We now classify AW*-algebras as:

I_{fin} = Type I, finite,

I_{inf} = Type I, purely infinite,

II_{fin} = Type II, finite,

II_{inf} = Type II, purely infinite,

III = Type III.

From the point of view of regular ring theory, the most interesting of these are the types I_{fin} and II_{fin}, for these are the algebras that have regular rings associated with them.

An AW*-algebra is a factor if 0 and 1 are the only central idempotents. In the finite case, the algebra is a simple ring, and its associated regular ring is a simple self-injective (right and left) ring.

Von Neumann also studied the lattice of projections of the type II_{fin} factors (the ordering of projections given earlier gives a lattice structure). These are modular lattices which are complete and have a kind of infinite modular law associated with them. Most of the algebraic abstraction that can be done at the ring theory level can be translated over to lattice theory and conversely. The two methods can also be usefully applied hand-in-hand.

§4. The Regular Ring of an Operator Algebra.

We have mentioned the regular ring of a finite von Neumann algebra. We shall now take a closer look at its construction and properties. Since the construction is basically algebraic, it is reasonable to look at the more general case of finite AW*-algebras.

Let $e_1 \leq e_2 \leq \ldots$ be a sequence of projections in a finite AW*-algebra A such that $\vee e_n = 1$. Berberian called such a sequence a strongly dense domain of A. We now consider a sequence of ordered pairs (x_n, e_n) where $x_n \in A$ and (e_n) is a strongly dense domain and such that $m<n$ implies $x_n e_m = x_n e_m$ and $x_n{}^* e_m = x_m{}^* e_m$. Berberian then defined an equivalence between those domains; namely $(x_n, e_n) \approx (y_n, f_n)$ if there is a

strongly dense domain (g_n) such that $x_n g_n = y_n g_n$ and $x_n^* g_n = y_n^* g_n$ for all n. Let us denote the equivalence class containing (x_n, e_n) by $[x_n, e_n]$. As he showed, one can define a multiplication and addition on the equivalence classes to obtain a ring R. We define addition on R by

$$[x_n, e_n] + [y_n, f_n] = [x_n + y_n, e_n \wedge f_n]$$

The definition of multiplication is more difficult and we refer the reader to [1].

Theorem [1]. The ring R constructed above with respect to the finite AW*-algebra A has the following properties -

a) A is a subalgebra of R,

b) R is a regular ring.

c) The involution on A extends to R and all the projections of R lie in A. Thus R is *-regular.

We note, for future reference, that in his proof of the above theorem, Berberian used the fact that for any element x in an AW*-algebra, $RP(x) \sim LP(x)$.

In [3] Berberian extended the construction of the regular ring to finite Baer*-rings satisfying certain additional axioms.

In 1951, Johnson defined the maximal ring of right quotients of a nonsingular ring R (here denoted by $Q_{maxr}(R)$). A right ideal E of R is essential if $E \cap I \neq (0)$ for any nonzero right ideal I of R. Let (E, α) be an ordered pair where E is an essential right ideal and

$\alpha: E \to R$ is a module homomorphism. We define an equivalence

$$(E,\alpha) \approx (E',\alpha')$$

if $\alpha|_{E \cap E'} = \alpha'|_{E \cap E'}$. Denote the equivalence class by $[E,\alpha]$. The equivalence classes are given a ring structure S using addition and composition (multiplication) of functions. If we assume that every essential right ideal of R has (0) as its left annihilator, then there is an embedding of R into S given by $x \mapsto (R,x)$ (where x as a homomorphism stands for left multiplication by x). This is the ring $Q_{maxr}(R)$.

One can certainly notice a similarity between Berberian's construction and Johnson's construction. However, it was not until 1968 that Roos announced (without proof) the following.

Theorem [27]. If A is a finite AW*-algebra, then its regular ring is isomorphic to its maximal ring of right quotients.

Proofs of the above theorem can be found in [12], [14] and [26]. Roos' theorem was generalized in these papers to certain finite Baer*-rings.

Theorem [14]. Let A be a Baer*-ring in which $\mathrm{ann}_r(x) = 0$ implies xA is an essential right ideal. Then $R = Q_{maxr}(A) = Q_{maxl}A$, the involution on A extends to R, R contains all the projections of A, and R coordinatizes the projection lattice of A.

Let us next consider an application of regular rings to AW*-algebras.

Theorem [2]. If A is an AW*-algebra, then A_n, the ring of nxn matrices over A, is an AW*-algebra.

The problem quickly reduces to the case where A is type II finite. It is routine to prove that A_n is a C*-algebra, so it remains to prove the annihilator condition. Let $R = Q_{maxr}(A)$, (the regular ring of A), then A_n is a subring of R_n and R_n is a self-injective directly finite *-regular ring. If $\{x_i\}$ is a set of elements in A_n, then its right annihilator in R_n is generated by a projection of R_n. We then prove that any projections in R_n lies in A_n, and this completes the proof of the theorem.

§5. Classification Theorems.

A C*-algebra A is AF (approximately finite-dimensional) if it is the C*-closure of the union, R, of a countable ascending chain of finite dimensional C*-subalgebras of A. The ring R is a direct limit of semisimple rings and R determines A up to isomorphism. Thus the classification of AF C*-algebras amounts to the classification of direct limits of semisimple algebras over $\mathbb{C}$.

Let us suppose that R_i is a finite direct sum of matrix rings over $\mathbb{C}$ and suppose we have an embedding of R_i into R_{i+1} for each i. Put $R = \varinjlim R_i$, the direct limit of the R_i. Since K_0 is a covariant functor, there are induced maps

$$K_0(R_i) \to K_0(R_{i+1})$$

and $K_0(R)$ is the direct limit of this directed system of abelian groups. Elliott showed that R is uniquely determined by the directed abelian group $(K_0(R),[R])$.

Theorem [8]. Let R be a countable direct limit of finite direct sums of matrix rings over $\mathbb{C}$ and let A be the C*-closure of R. Then:

a) $(K_0(R),[R])\cong(K_0(A),[A])$ (as directed abelian groups),

b) R is determined by $(K_0(R),[R])$.

Elliott also showed that a directed abelian group is isomorphic to the K_0 of some countable direct limit of finite direct sums of matrix rings over $\mathbb{C}$ if and only if it is a countable ordered abelian group which is the direct limit of a sequence of finitely generated ordered abelian groups with simplical positive cones (when represented in $\mathbb{Z}^n$).

Here is an interesting example of a simple regular ring that is a limit of finite direct sums of matrix rings, but is not a limit of simple artinian rings.

Define $$R_n = M_{f(2n-1)}\mathbb{C} \times M_{f(2n)}\mathbb{C},$$

where $f(m)$ is the m'th Fibonacci number $(f(1) = f(2) = 1, f(m) = f(m-1)+f(m-2$ and we define maps

$$\phi_n = R_n \to R_{n+1}, \text{ sending}$$

$$A \times B \to \begin{pmatrix} A & 0 \\ 0 & B \end{pmatrix} \times \begin{pmatrix} A & 0 & 0 \\ 0 & B & 0 \\ 0 & 0 & B \end{pmatrix}$$

Schematically this map is representable by the matrix $\binom{11}{12}$, and the $K_0(R)$ is $\lim(\mathbb{Z}^2 \to \mathbb{Z}^2)$. As $\det A = 1$, $K_0(R)$ is free of rank 2; in fact $K_0(R) = \mathbb{Z}+g\mathbb{Z}$, where g is the Golden ratio $(g-1 = g^{-1})$, as a subgroup of the reals. Of course, if R were a direct limit of simple artinian rings, $K_0(R)$ would have to be of rank 1. A small change in this example amounting to changing $\binom{11}{12}$ to $\binom{11}{21}$, changes $K_0(R)$ to $\mathbb{Z}+\sqrt{2}\mathbb{Z}$.

In another direction Cuntz [5,6] has made inroads in the study of simple C*-algebras, by associating to them a variant of K_0 where positive elements in the C*-algebra replace the projections (projectives). This invariant is very useful in distinguishing isomorphism types.

§6. Finite Rickart C*-Algebras.

A Rickart C*-algebra A is a C*-algebra in which the right annihilator of an element is generated by a projection. A consequence of this condition is that the right annihilator of any countable subset of A is generated by a projection. Therefore a Rickart C*-algebra is a countable analogue of an AW*-algebra.

In an attempt to apply the algebraic methods to finite Rickart C*-algebras, our first objective is the construction of a regular ring for such an algebra. The maximal ring of quotients, which worked well for AW*-algebras, is no longer suitable; however, a modification of the construction, where, instead of using the filter of essential right ideals, we use the filter of countably generated right ideals, works.

Theorem [15]. Let A be a finite Rickart C*-algebra. Then there is a ring of quotients R of A with the following properties:

a) R is a subring of the maximal ring of quotients of A,

b) R is unit regular,

c) the involution on A extends to R,

d) all projections of R lie in A (thus the projection lattices of the two rings are the same).

In an AW*-algebra the lattice of projections is a continous geometry. In a Rickart C*-algebra we cannot expect the lattice of projections to be a continuous geometry, however, the lattice is an $\aleph_0$-continuous geometry. Thus the lattice satisfies the conditions:

a) every countable subset has a supremum,

and b) if $b_1<b_2<\ldots$ is a countable ascending chain, and a is any element, then

$$a\wedge(\vee b_i) = \vee(a\wedge b_i).$$

One of the applications of the regular ring of a finite Rickart C*-algebra is the proof of the equivalence of right and left projections for any element x, that is,

$$RP(x)\sim LP(x)$$

This was not an application of the regular ring in the case of an AW*-algebra, in fact, as we mentioned earlier, Berberian used the equivalence of right and left projections to prove the existence of the regular ring.

As a consequence of this result and results on the existence of pseudo-rank functions, is the theorem that simple homomorphic images of a finite Rickart C*-algebras are finite AW* factors.

The second step in attacking finite Rickart C*-algebras is the characterization of K_0 of such a ring as a directed abelian group. Now a finite Rickart C*-algebra and its regular ring have isomorphic Grothendieck groups hence, the structure of K_0 is determined by considering the regular ring.

Theorem [19]. Let A be a finite Rickart C*-algebra with regular ring R. Then $K_0(R)$ is a directed abelian group with the properties:

a) [R] is an order unit,

b) every countable bounded ascending chain has a supremum; thus, $(K_0(R),[R])$ is countably monotone complete,

c) for any four elements $a,b,c,d \in (K_0(R),[R])$ such that a and b are both greater than c and d, there exists an element x such that $a \geq x, b \geq x, x \geq c$, and $x \geq d$; thus, $(K_0(R),[R])$ has the interpolation property.

It turns out that the above conditions are rather strong, in fact they imply that the group is a subgroup of $C(X,\mathbb{R})$, the real valued continuous functions from a compact Hausdorff space X with the induced pointwise ordering.

This, in turn, allows us to prove:

Theorem [19]. In a finite Rickart C*-algebra, the intersection of the maximal (two-sided) ideals is zero.

Further application of directed abelian groups to the study of finite Rickart C*-algebras will appear in [11].

§7. References.

[1] S. Berberian, The regular ring of a finite AW*-algebra, Annals of Math. 65 (1957), 224-240.

[2] __________, NxN matrices over an AW*-algebra, Amer. J. Math. 80 (1958), 37-44.

[3] __________, The Regular ring of a finite Baer*-ring, J. Alg. 23 (1972), 35-65.

[4] __________, Baer*-Rings, Grundlehren, Band 195, Springer-Verlag, (1972), New York.

[5] J. Cuntz, The structure of multiplication and addition in simple C*-algebras, Math. Scand. 40 (1977), 215-233.

[6] _______, Dimension functions on simple C*-algebras (to appear).

[7] G. Ehrlich, Unit regular rings, Portugal Math. 27 (1968), 209-212.

[8] G. Elliott, On the classification of inductive limits of sequences of semisimple finite-dimensional algebras, J. Alq. 38 (1976), 29-44.

[9] K. Goodearl, Von Neumann Regular Rings, (to appear).

[10] K. Goodearl and D. Handelman, K_0 and rank functions of regular rings, J. of Pure and Applied Algebra, 7 (1976), 195-216.

[11] K. Goodearl, D. Handelman, and J. Lawrence, $\aleph_0$- continuous rings and affine functions on a Choquet simplex, (to appear).

[12] I. Hafner, The regular ring and the maximal ring of quotients of a finite Baer*-ring, Michigan Math. J. 21 (1974), 153-160.

[13] D. Handelman, Perspectivity and cancellation in regular rings, J. Alg. 48 (1977), 1-16.

[14] __________, Coordinatization applied to finite Baer*-rings, Trans. Amer. Math. Soc. 235 (1978), 1-34.

[15] __________, Finite Rickart C*-algebras and their properties, Advances in Math. (to appear).

[16] __________, Finite Rickart C*-algebras and their properties II, Advances in Math. (to appear).

[17] __________, Stable range in AW*-algebras, Proc. Amer. Math. Soc. (to appear).

[18] __________, K_0 of von Neumann and AF C*-algebras, Oxford Quart. J. (to appear).

[19] D. Handelman, D. Higgs and J. Lawrence, Directed abelian groups, $\aleph_0$-continuous rings and Rickart C*-algebras, (to appear).

[20] D. Handelman and J. Lawrence, Finite Rickart C*-algebras, Bull. Amer. Math. Soc. 84 (1978), 157-158.

[21] I. Kaplansky, Projections in Banach algebras, Annals of Math. 53 (1951), 235-249.

[22] __________, Any orthocomplemented complete modular lattice is a continuous geometry, Annals of Math. 61 (1955), 524-541.

[23] __________, Rings of Operators, Benjamin (1968), New York.

[24] F. Murray and J. von Neumann, On rings of operators, Annals of Math. 37 (1936), 116-165.

[25] J. von Neumann, On regular rings, Proc. Nat. Acad. Sci. 22 (1936), 707-713.

[26] E. Pyle, The regular ring and the maximal ring of quotients of a finite Baer*-ring, Trans. Amer. Math. Soc. 203 (1975), 201-213.

[27] J.-E. Roos, Sur l'anneau maximal de fractions des AW*-algèbres et des anneaux de Baer, C. R. Acad. Sci. Paris Ser A-B, 266 (1968), A120-A133.

Principal Ideal Theorems

by Melvin Hochster[1]

1. Introduction

Unless otherwise specificied, all rings are commutative, with identity. "Local ring" means Noetherian ring with a unique maximal ideal. We shall discuss several results, some known, others new, which generalize Krull's classical principal ideal theorem. Several of the results are known only when the ring contains a field, while having the status of conjectures in the general case. In §6 we discuss a conjecture (6.5) on the existence of certain non-commutative overrings of local rings which is intermediate between the existence of big Cohen-Macaulay modules and the homological consequences thereof.

Sufficiently many definitions have been given to make all the statements of results and all the motivation accessible to algebraists who are not specialists in commutative Noertherian rings. However, some of the proofs, notably in §4, require quite a bit more

[1]The author was supported in part by a grant from the National Science Foundation.

background.

Results announced here without proof are proved in $[Ho_7]$.

§§1,2,3 are largely expository, but the material in §4 beginning with Theorem (4.3) is new. §§5 and 6 consist primarily of statements of some of the results of $[Ho_7]$.

2. Background on dimension theory.

One of our main contentions is that the dimension theory of commutative Noetherian rings, even finitely generated K-algebras, where K is a field or $\mathbb{Z}$, is not yet adequately understood. (In fact, all our questions "live" essentially in these down-to-earth cases: the existence of possibly bizarre Noetherian rings is not the issue.)

We recall that $\dim R$, R any commutative ring, is the supremum of lengths r of chains of primes $P_r \supsetneq \cdots \supsetneq P_0$ of R (R is never regarded as a prime, while (0) is prime precisely when R is a domain). If we restrict attention to chains such that $P_r = P$, a particular given prime, the supremum is denoted height P or ht P. Note that ht $P = \dim R_P$ (R_P is the localization of R at P, $(R-P)^{-1}R$).

Krull's original principal ideal theorem [K] asserts:

(2.1) Theorem. _Let_ R _be a Noetherian ring and let_ P _be a minimal prime of a principal ideal_ xR. _Then_ ht $P \leq 1$.

(2.2) Corollary. <u>Let</u> R <u>be a Noetherian ring and let</u> P <u>be a minimal prime of an</u> n-<u>generator ideal</u> $(x_1,\ldots,x_n)R$. <u>Then</u> ht $P \leq n$.

The derivation of (2.2) from (2.1) is a not difficult exercise, and for this reason (2.2) and (2.1) are both referred to as "Krull's principal ideal theorem". The difficult part is to prove (2.1): this is the first really deep and convincing theorem in the theory of abstract Noetherian commutative rings.

It is not hard to show, moreover, that if ht P = n then P really is minimal over an n generator ideal. This is a tremendously useful characterization of height in the Noetherian case. [To see how badly this can fail, let X = Spec R be the ordered set (or topological space) of primes of any commutative ring R . Then there is a possibly different commutative ring S such that Spec S $\approx$ Spec R (as ordered sets and even as topological spaces) and such that in S , if J is the radical of a finitely generated ideal (i.e., if the primes not containing J form a quasi-compact open set) then J is the radical of a principal ideal. In particular, if R is Noetherian, then S has the same space of primes but every radical ideal is the radical of a principal ideal. This follows from the construction in [Ho_1].]

If (R,m) is local, the principal ideal theorem implies that dim R = ht m is the least integer n such that there exist $x_1,\ldots,x_n \in m$ such that for some t ,

$$m^t \subset (x_1,\ldots,x_n)R .$$

The elements $x_1,\dots,x_n$ are then referred to as a "system of parameters" (s.o.p.) for R.

(2.3) Remarks. If $x_1,\dots,x_n$ are any elements of a Noetherian ring R, we can ask whether there exists a homomorphism $h:R \to S$, S again Noetherian, such that $(x_1,\dots,x_n)S$ is a proper ideal of height r, where, when I is not necessarily prime, $\operatorname{ht} I$ denotes $\inf\{\operatorname{ht} P : I \subset P,\ P \text{ prime}\}$. We note that this happens if and only if there exists $h:R \to S$, S local, such that $h(x_1),\dots,h(x_n)$ is a s.o.p. for S [for if P is a minimal prime of $(x_1,\dots,x_n)S$ of height n we may replace S by S_P.]

We conclude this section with a digression on the geometric significance of "height".

Geometrically, height corresponds to codimension. To see the analogy, consider the case where R is a finitely generated algebra over $\mathbb{C}$ with no nonzero nilpotents: we refer to this as the "geometric" case. If we represent $R = \mathbb{C}[x_1,\dots,x_n]/(f_1,\dots,f_m)$, where $x_1,\dots,x_n$ are indeterminates and $f_1,\dots,f_m$ are polynomials, then the set $V = \{x \in \mathbb{C}^n : f_1(x) = \dots = f_m(x) = 0\}$ corresponds to the maximal ideal space of R, and R may be identified with the ring of functions $V \to \mathbb{C}$ obtained as restrictions of polynomial functions.

We occasionally assume in the sequel that the reader is familiar with this set-up. For each radical ideal I of R, we have a composite surjection

$$\mathbb{C}[x_1,\dots,x_n] \twoheadrightarrow R \twoheadrightarrow R/I$$

so that R/I corresponds to a variety $V' \subset V$. If P is a prime ideal in a domain R and R , R/P correspond to varieties V, V' then

$$\text{ht } P = \dim V - \dim V' = \text{codim}_V V' ,$$

when $\dim V$ is, in fact, the same as $\dim R$, but may also be interpreted as the complex topological dimension (half the real topological dimension) of V .

3. What is a principal ideal theorem?

We shall refer to theorems which generalize Krull's principal ideal theorem (2.1),(2.2) or, sometimes, major special cases of Krull's principal ideal theorem, as "principal ideal theorems", but we also use this term for certain other results of a "similar flavor". To make this idea more precise, we want to reformulate Kull's theorem.

Let $R = \mathbb{Z}[X_1,\ldots,X_n]$ where $X_1,\ldots,X_n$ are indeterminates, and let $I = (X_1,\ldots,X_n)R$. Let $h: R \to S$, where S is Noetherian, be any homomorphism. To give such an h is the same as to specify values $x_1,\ldots,x_n$ for $X_1,\ldots,X_n$ in S . Krull's theorem then asserts that if Q is a minimal prime of $IS = (x_1,\ldots,x_n)S$, $\text{ht} Q \leq n$, i.e., for any homomorphism $R \to S$ such that IS is a proper ideal, height $IS \leq \text{ht } I$. (Note that this assertion is nontrivial even when $I = (X)$ in $\mathbb{Z}[X]$.)

We shall refer to any theorem which makes assertions of the type that "for a given ideal I in a Noetherian ring R and for all homomorphisms $R \to S$, S Noetherian, such that IS is proper, $\operatorname{ht}(IS) \leq \operatorname{ht} I$" as a "principal ideal theorem".

Recall that a Noetherian ring R is "regular" if the following equivalent conditions hold:

1) for every maximal ideal m, the maximal ideal of R_m is generated by $\dim R_m$ elements.

2) For every prime ideal P, the maximal ideal of R_P is generated by $\dim R_P$ elements.

3) Every finitely generated R-module has finite projective dimension.

Geometrically, the local ring at a point of an algebraic variety over $\mathbb{C}$ is regular if and only if the variety is smooth ($\equiv$ nonsingular) at the point (i.e., the Jacobian criterion permits one to give a local isomorphism with some $\mathbb{C}^r$ by invoking the implicit function theorem to solve for all but r of the variables in the defining equations in terms of the remaining r).

If R is regular, the situation with respect to principal ideal theorems is remarkably simple, by virtue of:

(3.1) Theorem. <u>Let</u> P <u>be a minimal prime of a regular Noetherian ring</u> R, <u>let</u> $h: R \to S$, S <u>Noetherian, and let</u> Q <u>be a minimal prime of</u> PS. <u>Then</u> $\operatorname{ht} Q \leq \operatorname{ht} P$.

This follows easily from results of Serre [S], p. V-18, Th. 3, and is proved in detail in [Ho_5], Theorem (7.1).

Note that (3.1) contains the reformulated Krull principal ideal theorem, since we may take $R = \mathbb{Z}[X_1,\ldots,X_n]$, which is regular, and $P = (X_1,\ldots,X_n)R$.

Hence, our main interest in the sequel is in the case where R is not regular. Very little is known.

(3.2) Example. Let K be a field, let $T = K[V,Y,U,V]$, let $R = T/(XY - UV) = K[x,y,u,v]$, and let $P = (x,u)$. Then $\dim R = 3$, while $R/P \cong k[Y,V]$, a polynomial ring, so that $\dim R/P = 2$ and $\operatorname{ht} P = 1$. Let $S = R/(y,v) = k[X,U]$, a polynomial ring in two variables. Then $PS = (X,U)$ has height 2, i.e., $\operatorname{ht} PS > \operatorname{ht} P$.

We are led by this example to formulate the following general question: let K be a field, the integers, or a complete discrete valuation ring such that p generates the maximal ideal, where p is a positive prime integer, let $X_1,\ldots,X_n,Y_1,\ldots,Y_q$ be indeterminates over K, and let T denote $K[X_1,\ldots,X_n, Y_1,\ldots,Y_q]$ or $K[[X_1,\ldots,X_n,Y_1,\ldots,Y_q]]$. ([[ ]] indicate formal power series.) In either case, T is regular. Let $F = F(X_1,\ldots, X_n,Y_1,\ldots,Y_q)$ be a nonzero element in $(X_1,\ldots,X_n)T$, let $R = T/(F)$, and let $P = (X_1,\ldots,X_n)R$. Clearly, P is a prime of R of height $n - 1$. (In Example (3.2), X_1,X_2 correspond to X,U, Y_1,Y_2 to Y,V, $F = XY - UV$, and $P = (X,U)R$.)

Now let y_1 be a point of $Y \cap Z$ and suppose $y_1 = g(y)$. Replacing f by fg we see that we may assume that $y = y_1$. Let Z^* be an irreducible component of $f^{-1}(Z)$ which maps onto Z. Then choose $z^* \in Z^*$ such that $f(z^*) = y$, and let Y^* be an irreducible component of $f^{-1}(Y)$ which maps onto Y. Since S is smooth, it follows from intersection theory in the smooth case (e.g., [S], p. V-18, Th. 3) that

$$\dim_{z^*}(Y^* \cap Z^*) \geq \dim Y^* + \dim Z^* - \dim S \quad .$$

Let W^* be an irreducible component of $Y^* \cap Z^*$ through z^* whose dimension is equal to that of $\dim_{z^*}(Y^* \cap Z^*)$. Then $f(W^*) \subset W$, whence

$$\begin{aligned} \dim W &\geq \dim f(W^*) \\ &= \dim W^* - \min\{\dim f^{-1}(t) \cap W^* : t \in f(W^*)\} \ . \end{aligned}$$

Since $\dim W^* = \dim_{z^*}(Y^* \cap Z^*)$ and since we may choose $t = f(z^*) = y$, we have

$$\dim W \geq \dim_{z^*}(Y^* \cap Z^*) - \dim f^{-1}(y) \cap W^* \ .$$

We already know $\dim(Y^* \cap Z^*) \geq \dim Y^* + \dim Z^* - \dim S$ while $W^* \subset Y^*$ implies $\dim f^{-1}(y) \cap W^* \leq \dim f^{-1}(y) \cap Y^*$. Thus,

$$\dim W \geq \dim Y^* + \dim Z^* - \dim S - \dim f^{-1}(y) \cap Y^* \ .$$

Since $\dim Y^* - \dim f^{-1}(y) \cap Y^* = \dim Y$, $\dim Z^* \geq \dim Z$, and $\dim S = \dim X$, this implies

$$\dim W \geq \dim Y + \dim Z - \dim X$$

or $t \geq r + s - n$ or $s - t \leq n - r$. Q.E.D.

Note that if $R = \mathbb{C}[X_1,\ldots,X_n]$, $P = (X_1,\ldots,X_n)$, and G is trivial, this theorem becomes Krull's principal ideal theorem for finitely generated $\mathbb{C}$-algebras.

We next prove a result which can sometimes be applied in situations resembling that of Example (3.3).

Let T be a regular local ring whose completion has the form $K[[x_1,\ldots,x_q]]$, where K is a field or complete discrete valuation ring, and suppose $R = T/(f)$, $f \neq 0$. Let $\dim R = n$. Given any finitely generated R-module $M = M_0$, we can form a minimal free resolution of M

$$\ldots \to F_1 \to F_0 \to M_0 \to 0$$

where the "minimality" means, equivalently that f_i maps onto a minimal basis for its image M_i , $i \geq 0$, or that $\operatorname{Im} F_{i+1} \subset mF_i$, where m is the maximal ideal of R , for $i \geq 0$.

Note that we have exact sequences

$$0 \to M_{i+1} \to F_i \to M_i \to 0 , \; i \geq 0 .$$

The M_i are determined uniquely up to nonunique isomrophism and are referred to as the "modules of syzygies" of $M = M_0$.

So far what we have said is valid for any Noetherian local ring R . In the case where R has the form $T/(f)$ described above, the results of [Ei] given much more: if $i > \dim R$, $M_{i+2} \cong M_i$ (this is true even for $i = \dim R$, provided M_i has no free direct summand).

In this case we denote by Ev(M) (resp. Od(M)) a module isomorphic to M_i for all sufficiently large even (resp. odd) integers i .

(4.4) Theorem. _Let_ (T,n) _be an unramified regular local ring and let_ $f \in n$, $f \neq 0$. _Let_ $R = T/(f)$ _and let_ P _be a prime of_ R . _Suppose that_ $\mathrm{Od}(R/P) \cong \mathrm{Ev}(R/P)$. _Then for every finitely generated_ R-_algebra_ S , _if_ $PS \neq S$ _then_ $\mathrm{ht}\, PS \leq \mathrm{ht}\, P$.

Proof. By localizing S at a suitable minimal prime of PS and killing a suitable minimal prime, we may assume that S is a local domain essentially of finite type over R . Replacing R by a localization of a polynomial ring over R (and P by P^e) , we may assume that $R \to S$ is surjective, with kernel $P' \in \mathrm{Spec}\, R$.

Thus, we reduce to the case where $S = R/P'$ and PS is primary to the maximal ideal of S , i.e., $P + P'$ is primary to the maximal ideal of R . What we need to show then is that $\mathrm{ht}\, P + \mathrm{ht}\, P' \geq \dim R$. This is immediate from the following:

(4.5) Theorem. Let (T,m) be an unramified regular local ring, $f \in n$, $f \neq 0$ and $(R,m) = (T'/(f), n/(f))$. Let P be a prime of R, and P' another prime such that $P + P'$ is primary to m. Assume $P, P' \neq m$. Then $\text{ht } P + \text{ht } P' \geq \dim R - 1$, and equality holds if and only if

$$\ell(H_m^0((Ev(R/P) \underset{R}{\otimes} P')) > \ell(H_m^0((Od(R/P) \underset{R}{\otimes} P'))$$

[Hence, if $Ev(R/P) \cong Od(R/P)$, $\text{ht } P + \text{ht } P' \geq \dim R$.]

Note: ℓ denotes length, and $H_m^0(E) = \bigcup_t \{u \in E : m^t u = 0\}$.

Proof. Let Q, Q' respectively be the inverse images of P, P' in T. By [S], p. V-15, Th. 2, $\text{ht } Q + \text{ht } Q' \geq \dim T$ with equality if and only if

$$\chi_T(T/Q, T/Q') > 0 .$$

Here, $\chi_T(E_1, E_2) = \Sigma(-1)^i \ell(\text{Tor}_i(E_1, E_2))$ (defined when T is regular local, E_1, E_2 are finitely generated, and $E_1 \otimes E_2$ has finite length) is the intersection multiplicity.

This translates at once to the given statement provided we can show that

$$\chi_T(T(Q, T/Q') = \ell^*_{ev}(P, P') - \ell^*_{od}(P, P') ,$$

where $\ell^*_{ev}(P,N) = \ell(H^0_m(Ev(R/P) \otimes_R N))$ and $\ell^*_{od}(P,N) = \ell(H^0_m(Od(R/P) \otimes_R N))$.

Let m be any large even integer, say bigger than $\dim T$. There is a long exact sequence [S], p. V-17

$$\cdots \to \mathrm{Tor}_{i-1}^R(T/Q,T/Q') \to \mathrm{Tor}_i^T(T/Q,T/Q') \to \mathrm{Tor}_i^R(T/Q,T/Q') \to \mathrm{Tor}_{i-2}^R(T/Q,T/Q') \to \cdots$$

which we may cut off at the $\mathrm{Tor}_m^T(T/Q,T/Q')$ term, (this is 0) and at the $\mathrm{Tor}_0^R(T/Q,T/Q')$ term. The alternating sum of the lengths of the terms in the truncated long exact sequence is 0. It readily follows that

$$\chi_T(T/Q,T/Q') = \ell(\mathrm{Tor}_m^R(T/Q,T/Q')) - \ell(\mathrm{Tor}_{n-1}^R(T/Q,T/Q'))$$

(each $\ell(\mathrm{Tor}_j^R(T/Q,T/Q'))$ term for $0 \le j \le n-2$ occurs twice, with opposite signs). Of course $T/Q = R/P$, $T/Q' = R/P'$.

Let M_i be the ith module of syzygies of R/P. Then for $h \ge 1$

$$\begin{aligned}\mathrm{Tor}_h^R(R/P,N) &\cong \mathrm{Tor}_{h-1}^R(M_1,N) \cong \cdots \cong \mathrm{Tor}_1^R(M_{h-1},N)\\ &\cong \mathrm{Ker}(M_h \otimes N \to F_{h-1} \otimes N)\end{aligned}$$

(from the short exact sequence $0 \to M_h \to F_{n-1} \to M_{h-1} \to 0$).
If N has no elements killed by a power of m except 0 and $R/P \otimes N$ has finite length (so that all the $\mathrm{Tor}_i^R(R/P,N)$ have

finite length), we can identify $\mathcal{T}or_h^R(R/P,N)$ with $H_m^0(M_h \otimes N)$ For h large and even, $M_h \cong Ev(R/P)$ and $M_{h-1} \cong Od(R/P)$. Q.E.D.

Remarks. It is possible to formulate various versions of these theorems in which R/P , R/P' are replaced by suitable modules. The condition that S be finitely generated over R can be relaxed. Moreover, what we really need in the regular rings in the proof is that their completions be formal power series rings over discrete valuation rings (rather than unramified, which is a somewhat stronger assumption).

If we consider the case where $R/(f) = R$ is itself regular, we recover Serre's Theorem 3, p. V 18 which is the basis for Theorem (3.1). In this case $Od(N) \cong Ev(N) \cong 0$ for every R-module N .

(4.6) Remark. The condition $Ev(R/P) \cong Od(R/P)$ can be weakened substantially in the normal case. To this end, we introduce an abelian group $\mathcal{H}(R)$ associated with any Noetherian normal domain R . Let $\mathcal{T}(R)$ be the set of isomorphism classes of finitely generated torsion-free R-modules. Consider the free abelian group with $\mathcal{T}(R)$ as basis and kill the subgroup generated by the elements

$$M - M_1 - M_2$$

whenever $M \cong M_1 \oplus M_2$ and

$$M_1 - M_2$$

whenever there is an exact sequence

$$0 \to F \to M_1 \to M_2 \to 0$$

with F <u>free</u>. Call the quotient $\mathcal{H}(R)$. We have a set map

$$\mathcal{T}(R) \to \mathcal{H}(R)$$

which takes $M \to [M]$. (We have not distinguished notationally between a module and its isomorphism class.)

It follows that every [M] is equal to some [I] for an ideal I of R .

Let $\mathcal{T}_2(R)$ be the set of isomorphism classes of torsion-free modules of depth 2 . Let $\mathcal{F}$ be the abelian group of $\mathbb{Z}$-valued functions on $\mathcal{T}_2(R)$. Then there is a map $M \mapsto \theta_M$ from $\mathcal{T}(R) \to \mathcal{F}$ (as sets) given by

$$\theta_M(L) = \ell(H_m{}^0(M \otimes_R L)) .$$

(4.7) Theorem. <u>The map</u> $M \mapsto \theta_M$ <u>from</u> $\mathcal{T}(R) \to \mathcal{F}$ <u>factors</u>

$$\mathcal{T}(R) \to \mathcal{H}(R) \overset{\overline{\theta}}{\to} \mathcal{F} ,$$

<u>where</u> $\overline{\theta}$ <u>is a group homomorphism</u>.

Proof. This immediately reduces to showing that

1) $\theta_{M_1 \oplus M_2} = \theta_{M_1} + \theta_{M_2}$

2) if $0 \to F \to M_1 \to M_2 \to 0$ is exact then $\theta_{M_1} = \theta_{M_2}$ if F is free.

1) is clear, while 2) follows if we can show that if L is torsion-free of depth 2, $H_m^0(M_1 \otimes L) \cong H_m^0(M_2 \otimes L)$.

Since $\mathrm{Tor}_1^R(M_2,L)$ must be torsion while $F \otimes L$ is torsion-free

$$0 \to F \otimes L \to M_1 \otimes L \to M_2 \otimes L \to 0$$

is exact. Since depth $L \geq 2$ and F is free, either $F \otimes L = 0$ or depth $(F \otimes L) \geq 2$. In either case, $H_m^i(F \otimes L) = 0$, $i = 0,1$, whence

$$0 \to H_m^0(M_1 \otimes L) \to H_m^0(M_2 \otimes L) \to 0$$

is exact. Q.E.D.

We thus obtain

(4.6) Theorem. Let (T,m) be an unramified regular local ring and let $f \in m$, $f \neq 0$. Let $R = T/(f)$ and suppose also that R is normal. Suppose that $\mathrm{Od}(R/P)$ and $\mathrm{Ev}(R/P)$ have the same class in $\mathcal{H}(R)$. Then for every finitely generated R-algebra

S , if $PS \neq S$ then $\text{ht } PS \leq \text{ht } P$.

Proof. We may make the same reduction as in the proof of Theorem (4.4). The hypothesis that $Ev(R/P)$ and $Od(R/P)$ have the same class in $H(R)$ is preserved upon adjoining indeterminates and localizing. (If S is flat over R , $[M] \to [S \otimes M]$ induces a map $H(R) \to H(S)$.) It remains to show that if P, P' are primes of R with $P + P'$ primary to m , then $\text{ht } P + \text{ht } P' \geq \dim R$. This is clear if $P = m$ or $P' = m$, so that we may assume otherwise. Then it suffices to show that

$$\ell(H_m^0(Ev(R/P) \otimes P')) = \ell(H_m^0(Od(R/P) \otimes P')) .$$

Since R is normal (and we may certainly assume $\dim R \geq 2$, or else R is regular), we have depth $R \geq 2$, and $P' \neq m \Rightarrow$ depth $R/P' \geq 1$ and so depth P' (as an R-module) is ≥ 2 . But then since $Ev(R/P)$, $Od(R/P)$ have the same image in $H(R)$ they have the same image in F , and the equality follows. Q.E.D.

Note that $H(R)$ is related to, but quite different from, Graham Evans' construction [Ev].

If R is a Dedekind domain, $H(R) \cong \text{Pic} R$, the ideal class group. If R is regular, $\dim R \leq 2$, and every projective is free, $H(R) = 0$. The study of $H(R)$ in general seems to be a difficult and important problem.

(4.7) Remark. If we look at the basic equivalence relations used in defining $\mathcal{H}(M)$, we see that if $\mathrm{syz}^1 M$ is any module obtained from M as the kernel

$$0 \to \mathrm{syz}^1 M \to F \overset{\alpha}{\to} M \to 0$$

from mapping a free module into M , then $[\mathrm{syz}^1 M]$ is independent of α and there is an endomorphism $\phi: \mathcal{H}(R) \to \mathcal{H}(R)$ such that

$$\phi([M]) = [\mathrm{syz}^1 M] .$$

If $R = T/(f)$, T regular and $n = \dim R$, then (see [Ei]) $\phi^q(\mathcal{H}(R))$ for $q \geq n-1$ is the subgroup of $\mathcal{H}(R)$ generated by the classes of the Cohen-Macaulay modules (depth n): call this $\mathcal{H}_\infty(R)$. ϕ is then an involution of $\mathcal{H}_\infty(R)$. The condition that $\mathrm{Od}(R/P) \cong \mathrm{Ev}(R/P)$ in Theorem (4.6) is then equivalent to the assertion that $\phi^q([R/P])$ is independent of (the parity of) q for large q .

5. Remarks on the direct summand conjecture.

We simply want to state here some partial results on the direct summand conjecture (3.4) discussed earlier in §3. Details will appear in [Ho_7].

A noetherian ring R is "supernormal" if it satisfies Serre's conditions R_2 and S_3 . This means that 1) if P is a prime of height ≤ 2 , R_P is regular; and 2) for any prime P , depth R_P

is at least $\min\{\text{ht } P, 3\}$. Thus if $\text{ht } P \leq 3$, R_P is Cohen-Macaulay, while if $\text{ht } P \geq 3$, depth $R_P \geq 3$.

Cohen-Macaulay rings are always S_3 , and regular rings are always supernormal.

(5.1) Theorem. a) _If R is a supernormal Noetherian domain and S is a normal domain which is a module-finite extension of R , then S/R is a reflexive_ R-module.

b) _Hence, if the degree of the extension of fraction fields is 2 , then S/R is a rank one reflexive._

c) _If, moreover, S is factorial (respectively, locally factorial), then S/R is free (resp. projective) of rank one, and R is a direct summand of S ._

(5.2) Corollary. _If R is regular, S is a domain module-finite over R, and the degree of the extension of fraction fields is 2 , then R is a direct summand of S ._

(5.3) Example. Let $R = \mathbb{Z}[x,y,u,v,] = \mathbb{Z}[X,Y,U,V]/(f)$, where $f = X - (UY^2 + 4V)$. Then R is a UFD, since 2 is prime and localizing at 2 yields a polynomial ring over $\mathbb{Z}[1/2]$. Consider the ring $S = R[u,(x + y\sqrt{u})/2]$. Then S is integral over R , the degree of field extensions is 2 , but R is not a direct summand of S as an R-module. This situation is preserved if one localizes at the prime (z,x,y,u,v) of R . [See [Ho_7];

also $[R_2]$, where this example is utilized in a related but somewhat different context.] Note that R is Cohen-Macaulay, hence S_3 , but not R_2 . (The localization at $(2,y,x)$ is not regular.)

(5.4) Theorem. Let (R,m) be a regular local ring and $x_1,\ldots,x_n$ a regular system of parameters, i.e., a minimal set of generators for m . Let $S \supset R$ be a module-finite ring extension of R . Suppose that

$$x_1^t \cdots x_n^t = \Sigma y_i x_i^r$$

in S . Then $\frac{t}{r} \geq \frac{2}{n}$.

For a proof, see $[Ho_7]$. Note that this eliminates the possibility

$$x_1x_2x_3 = \Sigma_{i=1}^3 y_i x_i^2$$

when $n = 3$, but not

$$x_1^2 x_2^2 x_3^2 = \Sigma_{i=1}^3 y_i x_i^3 .$$

Also note that the possibility

$$x_1x_2x_3x_4 = \Sigma_{i=1}^4 y_i x_i^2$$

remains when $n = 4$.

6. Concerning the existence of certain possibly noncommutative overrings.

In [Ho_7] a conjecture is discussed, the "canonical element conjecture", which is implied by the existence of big Cohen-Macaulay modules [Ho_4] and which in turn implies the usual homological consequences of the existence of big Cohen-Macaulay modules (e.g., the "new" intersection conjecture [PS_2], [R_1] , hence, the original intersection conjecture [PS_1], Bass' question [B], [PS_1] , M. Auslander's zerodivisor conjecture, [A_1], [A_2], [PS_1] and also the direct summand conjecture [Ho_3] and the Eisenbud-Evans principal ideal conjecture [EE].

In some sense, all the conjectured theorems are principal ideal theorems.

Our objective in this section is simply to state the canonical element conjecture, and observe that it would be implied by a conjecture on the existence of certain not necessarily commutative overrings. Moreover, the existence of these overrings is implied by the existence of big Cohen-Macaulay modules. No proofs are given: the details will appear in [Ho_7] .

We now systematically go through the conjectures involved.

(6.1) Conjecture. *If R is a local ring, $x_1,\ldots,x_n$ a s.o.p., then there exists an R-module M (a so-called "big" Cohen-Macaulay* module) such that $x_1,\dots,x_n$ is a regular sequence on M, i.e.:

1) $(x_1,\dots,x_n)M \neq M$; and

2) for $0 \leq i \leq n-1$ $(x_1,\dots,x_i)M : x_{i+1}R = (x_1,\dots,x_n)M$.

P. Griffith [G] has shown this is equivalent to the following:

(6.2) Conjecture. If (A,m) is an (m-adically) complete regular local ring and R is a domain module-finite over A, then there exists a nonzero countably generated free A-module which is an R-module.

In fact, in (6.2) it suffices to consider the case where A is a formal power series ring over a field or discrete valuation ring in which the residual characteristic generates the maximal ideal. I.e., $A = K[[X_1,\dots,X_n]]$ K a field, or else $A = V[[X_1,\dots,X_{n-1}]]$, V a complete DVR whose fraction field has char. 0, whose residue class field has char. $p > 0$, and the maximal ideal of V is pV.

(6.1) and (6.2) are equivalent and known if R (modulo a prime of coheight $\dim R$) contains a field or if $\dim R \leq 2$.

We next want to state the canonical element conjecture. This has a large number of formulations. We shall mention just two. First recall that if $x_1,\dots,x_n$ is any sequence of elements of R, $K_\cdot(x_1,\ldots,x_n;R)$, the Koszul complex of R with respect to $x_1,\ldots,x_n$, denotes the free complex in which the ith module has generators indexed by the i element subsets of $\{1,\ldots,n\}$, say $U_{j_1\ldots j_i}$, where $1 \le j_1 < \ldots < j_i \le n$, and

$$d\,U_{j_1\ldots j_i} = \Sigma_{t=1}^{i} (-1)^{t-1} x_{j_t} U_{j_1\ldots\hat{j}_t\ldots j_i} \quad .$$

Here ^ indicates omission.

In good cases, e.g., if $x_1,\ldots,x_n$ are a regular sequence on R, the Koszul complex is acyclic. It always has augmentation $R/(x_1,\ldots,x_n)$, and so provides a free resolution of $R/(x_1,\ldots,x_n)$ if $x_1,\ldots,x_n$ is a regular sequence.

(6.3) Conjecture. Let (R,m) be a local ring with $\dim R = n$, let $x_1,\ldots,x_n$ be a s.o.p., and let $k = R/m$. Consider a free exact resolution $L_\cdot$ of k truncated at the nth spot:

$$0 \to \mathrm{syz}^n k \to F_{n-1} \to \ldots \to F_1 \to R \to k \to 0 .$$

Lift the map $RU_\emptyset \overset{\cong}{\to} R$ which takes $U_\emptyset \mapsto 1$ (and induces the canonical surjection $R/(x_1,\ldots,x_n) \to k$) to a map of complexes $\psi: K_\cdot \to L_\cdot$, since $L_\cdot$ is exact and $K_\cdot$ is free (where $K_\cdot = K_\cdot(x_1,\ldots,x_n;R)$). This yields a map $\psi_n: RU_{1,\ldots,n} \to \mathrm{syz}^n k$.

Then, regardless of how $x_1,\ldots,x_n$ and ψ are chosen,

$\psi_n(U_{1,\ldots,n}) \neq 0$.

Remark. This is the most down-to-earth form of the conjecture. A simpler statement which is equivalent (see [Ho_7]) is:

(6.4) If (R,m) is local, with $k = R/m$, there exists an R-module M such that the map

$$\mathrm{Ext}^n(k,M) \to H_m{}^n(M) \quad \text{(local cohomology)}$$

is not zero.

Our algebra conjecture can now be stated as follows:

(6.5) Conjecture. Let $A = k[[X_1,\ldots,X_n]]$ or $V[[x_1,\ldots,x_{n-1}]]$, where k is a field and V is a complete discrete valuation ring of residual characteristic p with maximal ideal field pV . Let R be a domain module-finite over A . Then there exists a not necessarily commutative ring $N \supset R$ such that A (but not necessarily R) is in the center of N , an integer $q \geq 2$, and an endomrophism $F: N \to N$ over K (resp. V) such that:

[0) If the coefficient ring is V, N is V-flat.]

1) $F(X_i) = X_i{}^q$, $i \geq 1$.

2) If J is the ideal generated by the X's (resp. by the X's and p) , (X) is the ideal generated by the X's, $K.$ is the Koszul complex of N with respect to $X_1,\ldots,{}_n$ (resp. p,

$X_1,\ldots,X_{n-1})$, <u>and</u> $Z_i = \mathrm{Ker}(K_i \to K_{i-1})$, <u>then for sufficiently large</u> t

$$(\underline{X})^t K_i \cap Z_i \subset JZ_i \ , \ i \geq 1 \ .$$

[3) <u>If the coefficient ring is</u> V , $\overline{K}_1 = K_1/pK_1$, <u>and</u> $\overline{Z}_1 = Z_1/(pK_1 \cap Z_1)$, <u>then</u>

$$(\underline{X})^t \overline{K}_1 \cap \overline{Z}_1 \subset (\underline{X})\overline{Z}_1$$

<u>for all sufficiently large</u> t .]

4) $H_{(\underline{X})}{}^n(N) \neq 0$ (<u>resp</u>. $H_{(\underline{X})}{}^{n-1}(N/pN) \neq 0$) .

Conditions 2) and 3) assert that a weak form of the Artin-Rees theorem holds for K_i and its submodule Z_i (and $\overline{K}_1$ and $\overline{Z}_1$) for $i \geq 1$. This is a sort of weak "coherence" assertion for the ring N relative to the elements X_i .

One then has:

(6.6) Theorem. <u>The following implications are valid:</u>

$$\begin{array}{l} (6.1) \iff (6.2) \\ \quad \Downarrow \\ (6.5) \\ \quad \Downarrow \\ (6.3) \iff (6.4) \end{array}$$

The proof will be given in [Ho_7]. Note that (6.5) is <u>trivial</u> in char. $p > 0$, for one may choose $N = R$ and F = the Frobenius homomorphism. Thus, the implication (6.5) => (6.3) together with the fact that (6.3) implies the usual homological consequences of big Cohen-Macaulay modules yield a new proof of these consequences in char. $p > 0$. See [Ho_7] for more details.

References

[A_1] M. Auslander, Modules over unramified regular local rings, Illinois J. Math. 5 (1961), 631-645.

[A_2] M. Auslander, Modules over unramified regular rings, Proc. Intern. Congress of Math., 1962, 230-233.

[B] H. Bass, On the ubiquity of Gorenstein rings, Math. Z. 82 (1963), 8-28.

[Ei] D. Eisenbud, Homological algebra on a complete intersection, with an application to group representations, preprint, (Brandeis University).

[EE] D. Eisenbud and E.G. Evans, A generalized principal ideal theorem, Nagoya Math. J. 62 (1976), 41-53.

[Ev] E.G. Evans, Bourbaki's theorem and algebraic K-theory, J. of Algebra 41 (1976), 108-195.

[G] P. Griffith, A representation theorem for complete local rings, J. Pure and Applied Algebra 7 (1976), 303-315.

[Hi] H. Hironaka, Resolution of singularities of an algebraic variety over a field of characteristic 0 , Annals of Math. 79 (1964), 205-326.

[Ho_1] M. Hochster, Prime ideal structure in commutative rings, Trans. Amer. Math. Soc. 142 (1969), 43-60.

[Ho_2] M. Hochster, Cohen-Macaulay modules, Proc. Kansas Commutative Algebra Comference, Lecture Notes in Math., No. 311, Springer-Verlag, Berlin, Heidelberg, New York, 1973, 120-152.

[Ho_3] M. Hochster, Contracted ideals from integral extensions of regular rings, Nagoya Math. J. 51 (1973), 25-43.

[Ho_4] M. Hochster, "Topics in the Homological Theory of Modules over Commutative Rings", C.B.M.S. Regional Conference Series in Math., No. 24, Amer. Math. Soc., Providence, RI, 1974.

[Ho_5] M. Hochster, Big Cohen-Macaulay modules and algebras and embeddability in rings of Witt vectors, Proc. of the Queen's Univ. Commutative Algebra Conference (Kingston, Ontario, Canada, 1975), Queen's Papers in Pure and Applied Math., No. 42, 1975, 106-195.

[Ho_6] M. Hochster, Some applications of the Frobenius in characteristic 0 , Bull. Amer. Math. Soc., to appear.

[Ho_7] M. Hochster, Canonical elements in local cohomology modules, in preparation.

[K] W. Krull, Primidealketten in allgemeinen Ringbereichen, S.-B. Heidelbergen Akad. Wiss. Math.-Natur. Kl. (1928), 7.

[PS_1] C. Peskine and L. Szpiro, Dimension projective et cohomologie locale, Publ. Math. I.H.E,S., Paris, No. 42,(1973),323-295.

[PS_2] C. Peskine and L. Szpiro, Syzygies et multiplicités, C.R. Acad. Sci. Paris, Sér. A 278 (1974), 1421-1424.

[R_1] P. Roberts, Two applications of dualizing complexes over local rings, Ann. Sci. Ec. Norm. Sup. (4) 9, (1976), 103-106.

[R_2] P. Roberts, Abelian extensions of regular local rings, preprint.

[S] J.-P. Serre, "Algèbre Locale. Multiplicités." Lecture Notes in Math., No. 11, Springer-Verlag, Berlin, Heidelberg, New York, 1965.

[Sh] I.R. Shafarevich, "Basic Algebraic Geometry", Springer-Verlag, Berlin, Heidelberg, New York, 1977.

University of Michigan
Ann Arbor, Michigan 48109

Modules over the Cyclic Group of Prime Order

Lawrence S. Levy
Mathematics Department
University of Wisconsin
Madison, WI 53711

Abstract. We describe the structure of all finitely enerated modules over the integral group ring ZG, G = ⟨g⟩ cyclic of rime order p. The additive groups of the modules in question need not e torsion free. We give a moderately detailed description of the indecom-osable ZG-modules, and determine when two direct sums of such modules re isomorphic to each other.

Indications of proofs will be included, but full details will ppear elsewhere.

Note. Module will always mean finitely generated module unless he contrary is explicitly indicated. The notation G, g, and p used bove will be retained throughout this paper.

1. Separation into p-mixed and p-prime cases. A ZG-module vill be called p-mixed if its additive group is the direct sum of a torsion-free roup and a p-torsion group. It will be called p-prime if all of its elements ave finite order prime to p.

PROPOSITION 1.1. Every (finitely generated) ZG-module M can be written as a direct sum $M = M(p) \oplus M(p')$ of p-mixed and p-prime modules, respectively. Moreover, each summand is unique up to isomorphism.

Proof. $M(p')$ clearly has to be the set of all elements of M whose order is finite and prime to p . So define $M(p')$ by this condition. $M(p')$ is a ZG-submodule (in fact, a fully invariant subgroup) of M , and a Z-direct summand of M . Let $\pi : M \to M(p')$ be a Z-projection map. Since the elements of $M(p')$ can be uniquely "divided" by $p = |G|$, the map

$$m \to \frac{1}{|G|} \sum_{i=0}^{p-1} g^{-i} \pi g^{i}(m)$$

(stolen from one of the well-known proofs of Maschke's theorem) makes sense, and is a ZG-projection of M onto $M(p')$.

The complementary summand is clearly p-mixed and unique up to isomorphism, namely $\cong M/M(p')$.

The p-mixed case. This is the more interesting case, and will be discussed in §3, after the appropriate point of view has been introduced in §2

The p-prime case. The Krull-Schmidt theorem applies to direct sums of indecomposable p-prime modules because they have finite length. (In fact, they're finite.) Thus the only question to be answered is: What do the indecomposables look like?

As we shall see, in §4, these are objects we have all met before.

In §2 we introduce the basic notation of this paper. In §3 we discuss p-mixed modules in some detail.

Finally, in §4 we discuss the machinery which produced the results in §3;

2. Pullbacks and Separated Modules. Here we will describe the way of viewing the group ring ZG, in terms of which we will later describe all ZG-modules.

Given two group epimorphisms $f_i : S_i \to \bar{S}$ $(i = 1, 2)$ we will denote by $S = \{S_1 \xrightarrow{f_1} \bar{S} \xleftarrow{f_2} S_2\}$ the pullback of diagram 2 below, that is S is the set of symbols $(s_1 \to \bar{s} \leftarrow s_2)$ such that $f_1(s_1) = \bar{s} = f_2(s_2)$.

$$\begin{array}{ccc} Z = R_1 & & R_2 = Z[\theta] \\ & \searrow^{\nu_1} \quad {}^{\nu_2}\swarrow & \\ & \bar{R} = Z/(p) & \end{array}$$

Diagram 1.

$$\begin{array}{ccc} S_1 & & S_2 \\ & \searrow^{f_1} \quad {}^{f_2}\swarrow & \\ & \bar{S} & \end{array}$$

Diagram 2.

Note that our notation $(s_1 \to \bar{s} \leftarrow s_2)$ is a bit redundant; in fact $S \subseteq S_1 \oplus S_2$. (More precisely, the map $(s_1 \to \bar{s} \leftarrow s_2) \to (s_1, s_2)$ is a monomorphism.)

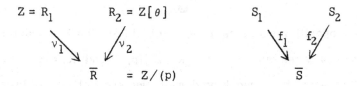

Next, let $Z[\theta]$ be the subring of the complex numbers generated by Z and a primitive p'th root θ of unity. Then we can construct the pullback $R = \{R_1 \xrightarrow{\nu_1} \bar{R} \xleftarrow{\nu_2} R_2\}$ shown in Diagram 1. Here ν_1 is the natural map, while ν_2 is the ring homomorphism which sends $1 \to \bar{1}$ and $\theta \to \bar{1}$. The starting point of our theory is the (known) observation:

PROPOSITION 2.1. $ZG \cong R = \{R_1 \xrightarrow{\nu_1} \bar{R} \xleftarrow{\nu_2} R_2\}$ via $g \to (1 \to \bar{1} \leftarrow \theta)$ and $1 \to (1 \to \bar{1} \leftarrow 1)$ (Ring $\cong$).

This proposition is a consequence of the fact that $ZG \cong Z[x]/(x^p-1)$, the factorization $x^p - 1 = (x-1)(x^{p-1} + \cdots + x + 1)$, and the fact that the polynomial $x^{p-1} + \cdots + x + 1$ is the minimal polynomial of θ over the rational numbers.

From now on, we will identify ZG and R via the isomorphism in 2.1.

The proposition suggests trying to obtain all $R = ZG$-modules as combinations of modules over $R_1 = Z$, $\bar{R} = Z/(p)$, and $R_2 = Z[\theta]$. Since R_2 is known to be a Dedekind domain, we know all of its finitely generated modules in a fair amount of detail.

Let S_i be an R_i-module $(i = 1,2)$ and $\bar{S}$ an $\bar{R}$-module. Note that $\bar{S}$ becomes an R_i-module $(i = 1,2)$ if we define $r_i\bar{s} = (\nu_i r_i)\bar{s}$. Suppose we are also given an R_i-linear map: $f_i : S_i \to \bar{S}$ $(i = 1,2)$. Then the pullback $S = \{S_1 \xrightarrow{f_1} \bar{S} \xleftarrow{f_2} S_2\}$ becomes an R-module if we define

$$(r_1 \to \bar{r} \leftarrow r_2)(s_1 \to \bar{s} \leftarrow s_2) = (r_1 s_1 \to \bar{r}\bar{s} \leftarrow r_2 s_2) .$$

What has to be checked is that the right-hand side belongs to S if the factors on the left belong to R and S respectively. And this follows from R_i-linearity of the f_i .

Any module S of this form will be called a separated R-module.

Separated R-modules form the second stage in a three-stage process for building all R-modules. The first stage consisted of modules over the coordinate rings R_1 , $\overline{R}$, R_2 . The third stage will be discussed in §§ 3 and 4 .

For clarity, we explicitly state the way in which G acts on a separated R-module S :

$$g \cdot (s_1 \rightarrow \overline{s} \leftarrow s_2) = (s_1, \overline{s}, \theta s_2)$$

(see Proposition 2.1) .

3. p-Mixed Modules. In order to be as explicit as possible, we note that $R_2 = Z[\theta]$ has exactly one prime ideal containing the prime number p . This prime ideal is generated by $p_2 = \theta - 1$, and $R_2/(p_2) \cong \overline{R}$ = the integers modulo p . (All of the notation of §1 will remain in force here.)

Stage 1. Our starting point for the construction of indecomposable, p-mixed $R = ZG$-modules will be the indecomposable p-mixed modules over R_1 , R_2 , and $\overline{R}$.

$R_2/(p_2^c)$ and ideals $\neq 0$ of R_2 ;

$R_1/(p^b)$ $[= Z/(p^b)]$ and $R_1 = Z$ itself.

Here b and c are arbitrary positive integers. Of course $\overline{R}$ is the only indecomposable $\overline{R}$-module.

<u>Stage 2</u>. We introduce a set of symbols $[b,c]$, where

c = a positive integer or an ideal $\neq 0$ of R_2

b = a positive integer or b = Z

(1) $$[b,c] = \{S_1 \xrightarrow{f_1} \overline{R} \xleftarrow{f_2} S_2\} \text{ where}$$

$S_2 = R_2/(p_2^c)$ if c is an integer; $S_2 = c$ if c is an ideal;

$S_1 = Z/(p^b)$ if b is an integer; $S_1 = Z$ if c = Z .

Note that epimorphisms: $S_i \to \overline{R}$ always exist: This is clear when b or c is an integer or when b = Z . When c is an ideal, use the formula $H/HK \cong R_2/K$ as R_2-modules (for H and K ideals $\neq 0$ in <u>any</u> Dedekind domain R_2) to obtain $f_2 : c \twoheadrightarrow c/c(p_2) \cong \overline{R}$.

<u>Stage 3</u>. Now choose an element of the form $s = (s_1 \to \overline{1} \leftarrow s_2) \in [b,c]$. Since (p_2) is the kernel of the map $R_2 \to \overline{R}$ which was used to define $R = \{R_1 \to \overline{R} \leftarrow R_2\}$ we see that $(0 \to 0 \leftarrow p_2)$ is an element of R . Hence, for any positive integer $c \geq 2$,

$$(0 \to 0 \leftarrow p_2)^{c-1} s = (0 \to \overline{0} \leftarrow p_2^{c-1} s_2) \in [b, c] .$$

Thus $[b, c]$ contains the R-submodule $\{0 \to 0 \leftarrow p_2^{c-1} S_2\} \cong \overline{R}$ which we call "the $\overline{R}$ of c ."

Similarly, when b is an integer ≥ 2 , $[b, c]$ contains the submodule $\{p S_1 \to 0 \leftarrow 0\} \cong \overline{R}$ which we call "the $\overline{R}$ of b ." We can therefore define the R-module

$$(2) \qquad [b_1, c_1] \smile [b_2, c_2] \smile \cdots \smile [b_n, c_n]$$

(each b_i and c_i an integer ≥ 2 except possibly for b_1 and c_n)

to be the direct sum $\oplus$ $[b_i, c_i]$ modulo the relation which identifies the $\overline{R}$ of c_i with the $\overline{R}$ of b_{i+1} $(1 \leq i < n)$.

THEOREM 3.1. Every R-module of the form (2) is indecomposable. If (2)' is another such module, then (2) $\cong$ (2)' $\iff$

(i) $n = n'$, every $b_i = b_i'$, $c_j = c_j'$ whenever $j < n$, and

(ii) $c_n = c_n'$ if c_n is an integer; $c_n \cong c_n'$ if c_n is an ideal.

Thus the particular homomorphisms used in (1) to form the building blocks $[b_i, c_i]$ turn out to be completely irrelevant (except that they must be onto)! This requires the fact that every unit of $\overline{R}$ can be lifted, via $R_2 \to \overline{R}$

to a unit of $R_2 = Z[\theta]$. This lifting property is not possessed by the map $R_1 = Z \to \bar{R}$, but it is, fortunately, not needed.

There is one other type of p-mixed indecomposable. If, in (2), b_1 and c_n are also integers ≥ 2 , then (2) will contain an $\bar{R}$ of b_1 and an $\bar{R}$ of c_n . Just identifying these with each other will sometimes produce an indecomposable module; and a more complicated identification involving b_1 and c_n and several other c_j's can also sometimes produce an indecomposable module. The details of this identification will be omitted. We will call these modules of type (3) , and we note, for emphasis,

(3) Modules of type (3) are obtained by imposing a single additional relation upon those of type (2) . Since every b_i and c_i is an integer, modules of type (3) are finite p-groups.

THEOREM 3.2. Every indecomposable p-mixed $R = ZG$-module is of type (2) or (3) . (Here we consider type (1) to be a special case of type (2)) .

Now we inquire when the following holds.

(4) $$M_1 \oplus \cdots \oplus M_s \cong M_1' \oplus \cdots \oplus M_t' \quad \text{(Each summand p-mixed and indecomposable.)}$$

THEOREM 3.3. Suppose (4) holds. If some M_i is of type (3), or of type (2) with its "c_n" an integer, then that $M_i \cong$ some M_j', and the isomorphism (4) still holds after M_i and M_j' have been cancelled.

To cover the situation remaining after all these cancellations, we set

(5) $$M_i = [b_{i1}, c_{i1}] \smile [b_{i2}, c_{i2}] \smile \cdots \smile [b_{i\,n(i)}, c_{i\,n(i)}] .$$

The most interesting part of the answer to our isomorphism question is given by:

THEOREM 3.4. Let modules $M_1, \cdots, M_s$ of the form (5) be given, with each $c_{i\,n(i)}$ an ideal; and let modules $M_1', \cdots, M_t'$ be given, of an analogous form (5)' with each $c_{i\,n'(i)}'$ an ideal.

If $M_1 \oplus \cdots \oplus M_s \cong M_1' \oplus \cdots \oplus M_t'$, then:

(i) The product of ideals $\prod_i c_{i\,n(i)}$ is in the same ideal class (in $Z[\theta]$) as the product $\prod_i c_{i\,n'(i)}'$; and

(ii) $s = t$; and, after suitable renumbering of the M_i, every $n(i) = n'(i)$, $b_{ij} = b_{ij}'$, $c_{ij} = c_{ij}'$ except that possibly $c_{i\,n(i)} \not\cong c_{i\,n'(i)}'$.

Conversely, if (i) and (ii) hold, then $\oplus M_i \cong \oplus M_i'$.

We close this section with some applications. First we ask how far from additively indecomposable can an indecomposable p-mixed ZG-module be?

THEOREM 3.5. Let M be a p-mixed, indecomposable ZG-module. Then

(i) The torsion-free rank of M must be 0, 1, p-1, or p .

(ii) The number of cyclic torsion summands can be unboundedly large in each of the cases enumerated in (i) . (In fact this can happen with the torsion subgroup of exponent 2 .)

Proof Sketch. For (i) , note that only modules of type (2) can have nonzero torsion-free parts, and only b_1 and c_n can contribute elements of infinite order. Then recall that, since the minimal polynomial of θ is $x^{p-1} + x^{p-2} + \cdots + x + 1$, $Z[\theta]$ and all of its ideals $\neq 0$ have torsion-free rank $p - 1$.

For (ii) it suffices to take

$$[b_1, 2] \smile [2, 2] \smile [2, 2] \smile \cdots \smile [2, c_n]$$

with $b_1 = 2$ or Z and $c_n = 2$ or $Z[\theta]$. QED.

In the structure of modules over a Dedekind domain, the separation into torsion and torsion-free cases is fundamental. For ZG-modules, this doesn't seem so useful. However, a clear-cut local-global separation holds.

DEFINITION. For a (finitely generated) p-mixed ZG-module $M = M_1 \oplus \cdots \oplus M_s$ (M_i indecomposable), let $c\ell(M)$, the <u>ideal class</u> of M in $Z[\theta]$, be the ideal class of the product $\prod_i c_{i\,n(i)}$ where the product is taken over all those i such that M_i has the form (5) above, with $c_{i\,n(i)}$ an ideal. (This is well-defined by Theorems 3.4 and 3.5.) If this product is empty we set $c\ell\, M$ = ideal class of $Z[\theta]$ itself.

Let Z_p be the p-localization of Z.

THEOREM 3.6. For p-mixed ZG-modules M and N, $M \cong N$ $(ZG \cong)$ $\iff$ $c\ell(M) = c\ell(N)$ and $M_p \cong N_p$ $(Z_pG \cong)$.

Proof sketch of the nontrivial direction ($\Longleftarrow$). Z_pG has a "pullback" description analogous to that of ZG, namely $Z_pG = \{Z_p \to \overline{R} \leftarrow Z_p[\theta]\}$. One first shows that, for M of type (2), $Z_p \otimes_Z M$ has a description of the form (2). This description is obtained by leaving all integers in (2) intact, replacing b_1 (if $b_1 = Z$) by Z_p, and replacing c_n (if c_n is an ideal) by $Z_p[\theta]$. If M is a p-mixed <u>torsion</u> module (in simpler language, a p-torsion group), then $Z_p \otimes_Z M \cong M$, and this takes care of modules of type (3).

Thus p-localization, while obliterating the specific ideals "c_n", still leaves a place-holder for each of them. Since we are given $c\ell(M)$ we merely fill these places with ideals whose product class equals that of $c\ell(M)$, and then use Theorem 3.4. QED.

4. Separated Representations. The purpose of this section is to present a very brief view of the machinery which produced the results of the previous section, and to briefly discuss p-prime modules. The notation of §1 remains in force here.

Basic Definition. A separated representation of an R-module M (R as in Proposition 2.1) is an R-module epimorphism

$$\varphi : S \twoheadrightarrow M$$

where S is a separated R-module

$$S = \{S_1 \xrightarrow{f_1} \overline{S} \xleftarrow{f_2} S_2\}$$

which is "as close as possible" to M in the sense that if φ has a factorization $\varphi : S \twoheadrightarrow S' \twoheadrightarrow M$ with S' also a separated R-module, then $S \twoheadrightarrow S'$ must be an isomorphism.

PROPOSITION 4.1. Every R-module M has a separated epresentation.

The proof is fairly straightforward. There certainly exist epimorphisms $g : F \twoheadrightarrow M$ with F separated, e.g. we can take F to be free. The set of submodules $K \subseteq \ker g$ such that F/K is separated has a maximal element K_1, and the map $F/K_1 \twoheadrightarrow M$ induced by g is easily seen to be a separated epresentation.

What makes separated representations useful is the following amazing property.

THEOREM 4.2. ("Almost Functorial Property"). Any homomorphism $f : M' \to M$ of R-modules can be lifted to a homomorphism of their separated representations. That is, if φ' and φ in the diagram below are separated representations, then there is a homomorphism f^* such that the diagram commutes.

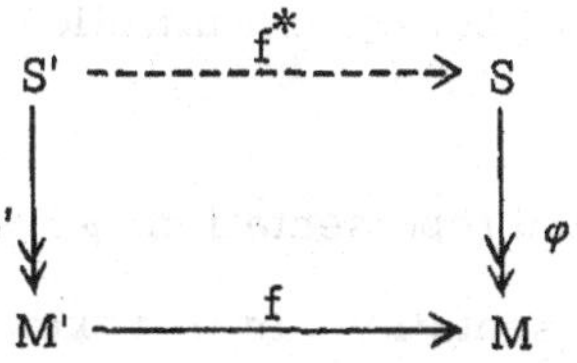

If f is 1 - 1 or onto, then so is any such f^*.

This is the main result of this investigation, from which everything else follows. Proposition 4.2 holds for R_1 and R_2 arbitrary rings, and $\overline{R}$ any semisimple artinian ring.

For one immediate consequence, take $M' = M$ and $f =$ the identity map. We get:

COROLLARY 4.3. (Uniqueness of Separated Representations). Let $\varphi_i : S_i \twoheadrightarrow M$ $(i = 1, 2)$ be separated representations. Then there is an isomorphism $f^* : S_1 \to S_2$ such that $\varphi_1 = \varphi_2 f^*$. In particular, $f^*(\ker \varphi_1) = \ker \varphi_2$.

The fact that makes separated representations useful in identifying indecomposable modules is: The direct sum of separated representations of modules M' and M is a separated representation of $M' \oplus M$.

THEOREM 4.4. Every indecomposable p-prime ZG-module M is either an indecomposable Z-module (g acting as multiplication by 1) or an indecomposable $Z[\theta]$-module (g acting as multiplication by θ).

Proof sketch. Take a separated representation $\varphi : S \twoheadrightarrow M$. One of the properties of separated representations is that $\ker \varphi$ is always an $\overline{R}$-module. In particular it is a p-group. On the other hand, since M is p-prime, S_1 and S_2 can also be shown to be p-prime. But

$$\ker \varphi \subseteq S \subseteq S_1 \oplus S_2 = \text{p-prime}$$

then forces $\ker \varphi = 0$. So $S \cong M$.

But $S = \{S_1 \to \overline{S} \leftarrow S_2\}$ with $\overline{S}$ an $\overline{R}$-module also makes $\overline{S} = 0$. So $S = \{S_1 \to 0 \leftarrow S_2\}$, which is just a fancy way of writing $S = S_1 \oplus S_2$. Indecomposability of M (hence S) finishes the job.

REMARKS on p-mixed modules. This case is much harder, but the strategy is the same. Take a separated representation $\varphi : S \twoheadrightarrow M$, and write $S = \{S_1 \xrightarrow{f_1} \overline{S} \xleftarrow{f_2} S_2\} \subseteq S_1 \oplus S_2$. View $K = \ker \varphi \subseteq S_1 \oplus S_2$ via its projection maps

$$S_1 \xleftarrow{\pi_1} K \xrightarrow{\pi_2} S_2 \ .$$

The task ahead is now clear: Find a "canonical form" for the four homomorphisms f_1, f_2, π_1, π_2.

Again the fact that $K = \ker \varphi$ is an $\overline{R}$-vector space helps out: we can represent π_1 and π_2 as matrices over $\overline{R}$. Moreover, if we factor f_i:

$$f_i : S_i \to \frac{S_i}{(\ker R_i \to \overline{R})S_i} = \overline{S}_i \xrightarrow{\overline{f}_i} \overline{S}$$

the fact that $\overline{S}_i$ is an $\overline{R}$-vector space makes it possible to represent $\overline{f}_1$ and $\overline{f}_2$ (hence, in effect, f_1 and f_2) by matrices over $\overline{R}$.

The resulting matrix reduction (rules provided by the Uniqueness Corollary 4.3) provides an interesting adventure in the simultaneous reduction of four matrices, involving simultaneous features of matrix similarity and matrix equivalence, and resulting in the results detailed in §3 .

REMARKS ON THE PROJECTIVE DIMENSION OF $\aleph$-UNIONS

Barbara L. Osofsky*

Rutgers University

New Brunswick, New Jersey 08903

The projective dimension of a module M (or object in an Abelian category) may be defined inductively as follows:

$$\mathrm{pd}(M) = 0 \quad \text{iff} \quad M \text{ is projective},$$

and for $0 < n < \infty$

$$\mathrm{pd}(M) = n \quad \text{iff} \quad \text{there exists a projective } P \text{ and a non-split exact sequence } 0 \to K \to P \to M \to 0 \text{ with } \mathrm{pd}(K) = n - 1.$$

If, for all $n \in \omega$, $\mathrm{pd}(M) \neq n$, we say $\mathrm{pd}(M) = \infty$.

General problem: Given M, compute its projective dimension.

Perhaps the greatest success in tackling this problem has been in the case of finitely generated modules over commutative Noetherian rings. The concept of projective dimension lies at the heart of the proof that any regular local ring is a UFD, and localization at a prime again gives a regular local ring. However, the situation to be discussed here is entirely different from this nice algebraic-geometric situation.

In certain cases, the projective dimension of a module can be shown to be determined completely by cardinality considerations. Let us look at some examples from the existing literature.

Let $R = F[x_1,\dots,x_n]$ be polynomials in n variables over a field F, and $M = F(x_1,\dots,x_n)$ the quotient field of R. Then pd(M) is the smaller of n and k + 1, where $\mathrm{card}(F) = \aleph_k$. (See [3] or [6].)

---

*Research partially supported by N.S.F. Grant MPS75-07580.

Let K be a separably generated field extension of F. Then, as a $K \otimes_F K$ -module, pd(K) = trans.deg.(K/F) + k + 1, where K is generated as a field extension of F by $\aleph_k$ but no fewer elements. (See [5].)

Let $\{ e_\alpha \mid \alpha \varepsilon I \}$ be a set of commuting idempotents in a ring R such that $(\Pi_{\alpha\varepsilon F}\, e_\alpha)\, (\Pi_{\beta\varepsilon G}(1 - e_\beta)) \neq 0$ for any pair of finite disjoint subsets F and G of I. Then $\mathrm{pd}(\Sigma_I\, e_\alpha R) = k$, where $\aleph_0 \cdot \mathrm{card}(I) = \aleph_k$. (See [8], [4], or [6].)

For additional examples, see [6] or [7].

In all of these cases except for the examples in [7], the problem of computing a projective dimension in terms of the subscript of an $\aleph$ is reduced to finding the dimension of some module which is a direct union of cyclic projective modules. The cyclic projective hypothesis is used to construct a projective resolution analagous to the Koszul complex of commutative algebra or the complex of chains in combinatorial topology. Different theorems require slightly different resolutions. The particular form of the resolution is used to get a lower bound on projective dimension. There seems no reason why such a specific resolution should be needed. As a matter of fact, it isn't. In [7], I get rid of the particular resolution by abstracting out what properties of it I need in the proofs. There are very few novel proofs in [7]--the newness is mostly in the definitions. Existing proofs are recast in terms of these new definitions, thereby making them much more powerful and applicable to additional cases.

In this paper, I will give the basic definitions of [7], and indicate how the major theorems of that paper can be used to compute a specific projective dimension (cf. [2]). In this application, both a typical projective resolution and the new concepts of [7] are used.

Let me start off by saying what kinds of modules we will be looking at, and what kind of a result we are after.

Problem: You have a module M which is a sum of submodules $\{N_\beta \mid \beta \in \Lambda\}$ and you wish to calculate $\mathrm{pd}(\Sigma_{\Lambda'} N_\beta)$ for $\Lambda' \subseteq \Lambda$. Find some "reasonable" hypotheses on the N_β's and how they fit together so that the projective dimension depends only on the cardinality of a minimal generating set for $\Sigma_{\Lambda'} N_\beta$. Specifically, we wish to prove a theorem of the sort:

$$(*) \quad \left\{ \begin{array}{l} \text{There is a } k \in \omega \text{ such that, for } \Lambda' \subseteq \Lambda, \\ \mathrm{pd}(\Sigma_{\Lambda'} N_\beta) \leq n + k \text{ iff } \Sigma_{\Lambda'} N_\beta \text{ is } \aleph_{\nu+n}\text{-generated,} \\ \text{where } \nu \text{ is an ordinal or } -1 \text{ (standing for "finite").} \end{array} \right.$$

We impose the following two conditions on $\{N_\beta \mid \beta \in \Lambda\}$.

(1) Each N_β should possess a projective resolution consisting of $\aleph_\nu$-generated projectives, that is, each N_β should be $\aleph_\nu$-resolvable. Some condition of this sort is necessary to insure that there is no growth in cardinality as you use finite induction to move down a projective resolution.

(2) Sums are messy, directed unions are somewhat less messy since the category of modules over a ring has exact direct limits. Hence we add a hypothesis that, "without loss of generality", $\{N_\beta \mid \beta \in \Lambda\}$ is directed by $\subseteq$. Specifically, if it is not, then there is some family $F \subseteq \{\Sigma_{i=1}^m N_{\beta_i} \mid m \in \omega, \beta_i \in \Lambda\}$ such that $F \cup \{N_\beta \mid \beta \in \Lambda\}$ is directed under $\subseteq$ and this new set still has property (1).

All proofs are by finite induction. One implication in (*) is rather easy to show provided we have the basis for the induction and the directed property of (2). By a straightforward application of a 1955 proposition due to Auslander [1], if $\mathrm{card}(\Lambda') \leq \aleph_{\nu+n}$, then $\mathrm{pd}(\Sigma_{\Lambda'} N_\beta) \leq n + k$ provided the result holds when $n = 0$ (i.e., $\aleph_\nu$-generated sums have dimension $\leq k$). There is nothing at all new about this proof--it is the one that has been used for the upper bound implication of (*) from the beginning. It is the other implication of (*) that is difficult.

In the general theory used to obtain the only if or lower bound implication of (*), namely,

$$\Sigma_{\Lambda'} N_\beta \text{ is not } \aleph_{\nu+n}\text{-generated implies}$$
$$\mathrm{pd}(\Sigma_{\Lambda'} N_\beta) > n + k \quad ,$$

a new concept is introduced. Call $\Lambda' \subseteq \Lambda$ nice if $\{N_\beta | \beta \varepsilon \Lambda'\}$ is directed.

$$\Delta_n = \{ \Sigma_{\Lambda'} N_\beta \mid \Lambda' \text{ nice}, \mathrm{card}(\Lambda') = \aleph_{\nu+n} \}$$

is studied for different $n \in \omega$. The properties of this family of significance for the investigation are summarized in the following definition.

<u>Definition</u>. Let $\aleph$ be an infinite cardinal, Δ a directed poset, M a module (or object in an Ab5 category), and

$$\{ M_\alpha \to M_\beta \mid \alpha < \beta \text{ in } \Delta \}$$

a directed system of subobjects of M. Then M is called the <u>$\aleph$-union</u> of $\{M_\alpha \mid \alpha \in \Delta\}$ provided:

i) $M = \bigcup_{\alpha\varepsilon\Delta} M_\alpha$,

ii) Each M_α has a projective resolution consisting of $\aleph$-generated projectives ($=_{\mathrm{def}}$ M_α is $\aleph$-resolvable) ,

iii) If $D \subseteq \Delta$ is directed and $\mathrm{card}(D) \leq \aleph$, then there is an $\alpha \in \Delta$ such that $\bigcup_{\beta\varepsilon D} M_\beta = M_\alpha$ and $\alpha \geq \beta$ for all $\beta \in D$.

Note that, for each n, $\bigcup\Delta_n$ is the $\aleph_{\nu+n}$-union of Δ_n. That is why the concept is useful for our problem.

By means of a series of small steps, most of which are known or within epsilon thereof, and the remainder of which are harder to formulate than to prove, one arrives at the main theorem about $\aleph$-unions. This theorem is at the heart of existing proofs calculating projective dimension in terms of cardinality. The proof uses the inductive definition given at the beginning of this paper to do finite induction on projective dimension.

heorem A. Let M be the $\aleph$-union of $\{ M_\alpha \mid \alpha \in \Delta \}$, M not $\aleph$-generated, and let Ω be the first ordinal with cardinality > $\aleph$. Assume $pd(M) \leq k < \infty$. Then there exists a set $D \subseteq \Delta$ such that D and $\{ M_\alpha \mid \alpha \in D \}$ have order type Ω and, for all $\alpha \in D$,

$$pd(M/\bigcup_{\beta<\alpha} M_\beta) \leq k \quad .$$

There is an interesting way to look at what Theorem A says. If $\aleph$ appens to be a regular cardinal, property iii) of the definition of $\aleph$ -union gives an $\alpha \in \Delta$ with $pd(M/M_\alpha) \leq k$ where M_α is generated by $\aleph$ ut no fewer elements. Call an M_α requiring $\aleph$ generators "big". ince $pd(M) \leq k$ and $pd(M/M_\alpha) \leq k$, standard results tell us that d$(M_\alpha) \leq k$. Now go the other way. If we know that $pd(M) \leq k$ and d$(M_\alpha) \leq k$, we know $pd(M/M_\alpha) \leq k + 1$. But M/M_α does not achieve the aximum projective dimension possible from this formula. Hence we may ay that M_α is "big" but $pd(M/M_\alpha)$ is "small". The problem then becomes ow to exploit this "big"-"small" relationship. With one exception, ll pre- $\aleph$ -union proofs employed (without saying so) the following.

heorem B. Let N, $L \subseteq M$, and $pd(M) \leq k$, $pd(M/N) \leq k$, $pd(N + L) \leq k$, and $pd(L) < k$. Then $pd(N \cap L) < k$.

roof. Apply well known relationships on projective dimensions to the xact sequences

$$\begin{array}{ccccccccc}
 & & 0 & & & & 0 & & \\
 & & \downarrow & & & & \downarrow & & \\
 & & N \cap L & & & & N + L & & \\
 & & \downarrow & & & & \downarrow & & \\
 & & L & & & & M & & \\
 & & \downarrow & & & & \downarrow & & \\
0 & \to & (N + L)/N & \to & M/N & \to & M/(N + L) & \to & 0 \\
 & & \downarrow & & & & \downarrow & & \\
 & & 0 & & & & 0 & &
\end{array}$$

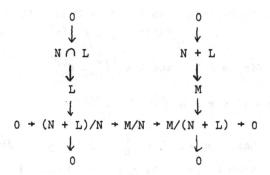

Now let us return to the problem of proving (*) using Theorems A nd B and finite induction. Let me summarize a theorem that works in almost all (*)-type theorems and then use it in an application modeled after Chen [2]. Other applications may be found in [7].

Since the hypotheses of Theorem C below seem rather strange, a few remarks are in order before the theorem is stated. The hypotheses are precisely what is needed to insure that finite induction will work, as the proof following the theorem shows. B) and C) are the basis for finite induction, and A) and D) say Theorems A and B provide the induction step. Note that the computation of an upper bound is used in the computation of the lower bound. For n>0, $\Delta_n^{(F)} - \Delta_{n-1}^{(F)}$ picks out precisely those modules generated by $\aleph_{\nu+n}$ but no smaller number of generators.

<u>Theorem C</u>. Let $M = \Sigma_{\beta\in\Lambda} N_\beta$. For all F and G finite subsets of Λ, set

$$N_F = \bigcap_{\beta\in F} N_\beta \quad ,$$

and assume $N_F \cap \Sigma_{\beta\in G} N_\beta$ is $\aleph_0 \cdot \aleph_\nu$-resolvable. For $n \in \omega$, set

$$\Delta_n^{(F)} = \{\ \Sigma_{\beta\in\Lambda'} N_F \cap N_\beta \mid \Lambda' \subseteq \Lambda,\ \mathrm{card}(\Lambda') \leq \aleph_{n+\nu}\ \}$$

and $\Delta_{-1}^{(F)} = \emptyset$. If, for all finite F and G and some $k < \infty$,

A) $N_F \cap \Sigma_{\beta\in G} N_\beta = \Sigma_{\beta\in G}(N_F \cap N_\beta)$,

B) For all $N \in \Delta_0^{(F)}$, $\mathrm{pd}(N) \leq k$,

C) For all N and n, if $N \in \Delta_n^{(F)}$ and $\mathrm{pd}(N) \leq k$, then $N \in \Delta_0^{(F)}$

D) If $N \in \Delta_n^{(F)} - \Delta_{n-1}^{(F)}$ and $\Lambda' \subseteq \Lambda$ has cardinality $> \aleph_{\nu+n}$, then there exists an $\alpha \in \Lambda' - F$ such that

$$N \cap N_\alpha \in \Delta_n^{(F\cup\{\alpha\})} - \Delta_{n-1}^{(F\cup\{\alpha\})} \quad ,$$

then for all finite $F \subseteq \Lambda$ and $N \in \bigcup_{n=0}^{\infty} \Delta_n^{(F)}$,

$$\mathrm{pd}(N) \leq k + n \iff N \in \Delta_n^{(F)}$$

<u>Proof</u>. B) states that, for all $N \in \Delta_0^{(F)}$, $\mathrm{pd}(N) \leq k$. Auslander's proposition in [1] then gives $\mathrm{pd}(N) \leq k + n$ for all $N \in \Delta_n^{(F)}$ as sketched above. This is the upper bound inequality of (*). In particular, every element of $\bigcup_{n=0}^{\infty} \Delta_n^{(F)}$ has finite projective dimension.

By C), if $pd(N) \leq k$, then $N \in \Delta_0^{(F)}$. Now assume that, for $n \in \omega$, $d(N') \leq k + n$ implies $N' \in \Delta_n^{(F')}$ for any finite F', and let pd(N) = $k + n + 1$ for $N \in \bigcup_{m=0}^{\infty} \Delta_m^{(F)}$. By the initial hypotheses, $\bigcup \Delta_{n+1}^{(F)}$ is the $\aleph_{\nu+n+1}$-union of $\Delta_{n+1}^{(F)}$. By Theorem A, if $N \notin \Delta_{n+1}^{(F)}$, there is an N' in $\Delta_{n+1}^{(F)} - \Delta_n^{(F)}$ with $pd(N/N') \leq k + n + 1$ (large kernel, small quotient). By D) and A), there is an α in the indexing set for N such that $N_\alpha \cap N'$ is in $\Delta_{n+1}^{(F \cup \{\alpha\})} - \Delta_n^{(F \cup \{\alpha\})}$, that is, $N_\alpha \cap N'$ stays big by D) and we get into $\Delta_{n+1}^{(F \cup \{\alpha\})}$ by A). But since $k < k + n + 1$, Theorem B says $pd(N_\alpha \cap N') \leq k + n$, contradicting the induction hypothesis. Hence $N \in \Delta_{n+1}^{(F)}$.

An Application

A theorem such as Theorem C would be rather meaningless if there were no natural situations to which it applied. I outline one application here, and refer you to [7] for others. The object is to compute the global dimension of a skew group ring of an Abelian group G. In [2], Chen computes that dimension for all G except for countable groups of torsion-free rank > 1 with finite uniform dimension. Theorem D below is the major portion of his computation.

Theorem D. (Chen [2]). Let $G = \oplus_{\alpha \in A} (\mathbb{Z}/n_\alpha \mathbb{Z})$, $n_\alpha \neq 1$, K a field, $\phi: G \to Aut(K)$ a group homomorphism. Let $R = K_\phi[G]$ be the skew group algebra, that is, (R,+) is the right vector space over K with basis G, and

$$\alpha \cdot g = g(g^\phi(\alpha)) \quad \text{for all} \quad \alpha \in K, g \in G,$$

where g^ϕ is the image of g under ϕ. If, for every finite subgroup $H \subseteq \ker \phi$, char(K) does not divide the order of H, then R contains a right ideal I with

$$pd(R/I) = m+h+1$$

where $card(A) = \aleph_m$ and $h = rank(G) = card\{\alpha \in A | n_\alpha = 0\}$.

Proof. We outline the proof here. Missing details may be found in [2], where the theorem was proved for ϕ 1-1.

The hypothesis on finite sugbroups is equivalent to the statement that, for any finite $H \subseteq G$, $K_\phi[H]$ is semi-simple. The author wishes to thank Robert Warfield for providing a proof of that. Specifically, let H be a finite subgroup of G, I an ideal of $K_\phi[H]$, $0 \neq x = \sum_{i=1}^m g_i \alpha_i \in I$ with minimal m. Let $\beta \in K$. Then

$$x-\beta^{-1}xg_m^\phi(\beta) = \sum_{i=1}^{m-1} g_i(1-g_i^\phi(\beta^{-1})g_m^\phi(\beta))\alpha_i$$

is of shorter length than x, and so $= 0$. Thus $g_i(\beta) = g_m(\beta)$ for all i and $\beta \in K$, so $g_m^{-1}x \in K_\phi[H \cap \ker \phi]$ which is semi-simple by Maschke's theorem. I is therefore not nilpotent.

Now let $n_1 = \ldots = n_h = 0$, $n_\beta \neq 0$ for all other β in A. Let g_α denote the element of G corresponding to $1+n_\alpha\mathbf{Z}$. Set

$$\Lambda = A-\{1,2,\ldots,h\}.$$

For $\beta \in \Lambda$, set

$$N_\beta = \sum_{i=1}^{h} (1-g_i)R+(1-g_\beta)R.$$

Set

$$I = \sum_{\beta\in\Lambda} N_\beta.$$

If G is not finite, $pd(R/I) = 1+pd(I)$. One checks that, for any finite F, $F' \subseteq \Lambda$,

$$N_F = \bigcap_{\beta \in F} N_\beta = \sum_{i=1}^{h} (1-g_i)R + \Pi_{\beta \in F}(1-g_\beta)R;$$

$$N_F \cap \sum_{\gamma \in F'} N_\gamma = \sum_{\gamma \in F'} (N_F \cap N_\gamma);$$

R is (right) coherent so finitely generated right ideals are $\aleph_{-1}$ resolvable;

If $L \subseteq \Lambda$ and $\alpha \notin L$, then $N_\alpha \cap N_F \cap \sum_{\beta \in L} N_\beta$ requires as many generators as $N_F \cap \sum_{\beta \in L} N_\beta$.

We thus have everything needed to apply Theorem C except hypotheses B) and C). We next consider two cases.

Case i) G is torsion, that is, $h = 0$. Let $\nu = 0$, $k = 0$ in Theorem C. Then R is a von Neumann regular ring and so countably generated ideals are projective, giving B). Now let $N = N_F \cap \sum_{\beta \in L} N_\beta$ be projective. If L is not countable, then by Theorem A, there is an infinite countable $L' \subseteq L$ with $N/\sum_{\beta \in L'} (N_F \cap N_\beta)$ projective. (Originally, Kaplansky's theorem that a projective module is a direct sum of countably generated submodules was quoted at this point. Theorem A is basically the result of a jazzed up version of Kaplansky's proof.) Let $N = \sum_{\beta \in L'} (N_F \cap N_\beta) \oplus M$, ρ project N to M. Then for $\alpha \in L-L'$, $\rho((1-g_\alpha)(\pi_{\beta \in F}(1-g_\beta))R)$ is projective but its kernel is not finitely generated, a contradiction.

Case ii) If $h > 0$, let $\nu = -1$, $k = h-1$ in Theorem C. Note that, for $1 \leq i \leq h$, $1-g_i$ is not a zero divisor in R. To show B) let F' be a finite subset of Λ,

$$N = N_F \cap \sum_{\gamma \in F'} N_\gamma,$$

$$N' = \sum_{\gamma \in F'} (\Pi_{\beta \in F}(1-g_\beta))(1-g_\gamma)R.$$

Then $(1-g_1)R \cap N' = (1-g_1)\cdot N'$, N' is a direct summand of R, and

$1-g_1$ is not a zero divisor. Thus $(1-g_1)N'$ is a direct summand of $(1-g_1)R$ and $(1-g_1)R+N'$ is projective. By induction on ℓ,

$$pd(\sum_{i=1}^{\ell} (1-g_i)R+N') \leq \ell-1.$$

Now let $N = N_F \cap \sum_{\beta \in L} N_\beta$ have $pd(N) \leq h-1$. To show C) of Theorem C, we must show L is finite. To do this, Chen constructs a specific projective resolution. The usual bar resolution in the case $G = \ker \phi$ does not seem to help, but this resolution bears some resemblance to it. The independence of $\{g_\alpha | \alpha \in A\}$ is necessary to show exactness and Abelian is necessary to show that we have a complex.

Set $p = \Pi_{\beta \in F}(1-g_\beta)$. Linearly order Λ' to keep signs consistent. For $\alpha_0 < \alpha_1 < \dots \alpha_n \in \Lambda'$, let $<\alpha_0,\dots,\alpha_n> \in R^{L^{n+1}}$ denote that function whose value at $(\alpha_0,\dots,\alpha_n)$ is $p \cdot \Pi_{j=0}^{n}(1-g_{\alpha_j})$ (or just $\Pi_{j=0}^{n}(1-g_{\alpha_j})$ if every $\alpha_j \in \{1,\dots,h\}$) and whose value elsewhere is 0. Let $P_n \subseteq R^{L^{n+1}}$ be generated by these functions $\{<\alpha_0,\dots,\alpha_n>\} \subseteq R^{L^{n+1}}$. Define $\partial_0 : P_0 \to N$ by

$$\partial_0<i> = (1-g_i) \qquad 1 \leq i \leq h$$

$$\partial_0<\alpha> = p\cdot(1-g_\alpha) \qquad \alpha \notin \{1,\dots,h\}$$

and for $n \geq 1$

$$\partial_n<\alpha_0,\dots,\alpha_n> = \sum_{j=0}^{n} <\alpha_0,\dots,\hat{\alpha}_j,\dots,\alpha_n>(-1)^j(1-g_{\alpha_j})$$

where $\hat{\alpha}_j$ means omit α_j. Then

$$\dots \to P_n \xrightarrow{\partial_n} P_{n-1} \to \dots \to P_0 \xrightarrow{P_0} N \to 0$$

is a projective resolution of N. Since $\partial_{h-1}P_{h-1}$ is projective, the exact sequence

$$0 \to \partial_{h-1}\langle 1,\ldots,h\rangle R \cap A \to \partial_{h-1}\langle 1,\ldots,h\rangle \oplus A \to \partial_{h-1}P_{h-1} \to 0$$

splits, where $A = \sum \langle \alpha_0,\ldots,\alpha_{h-1}\rangle R$ with some $\alpha_i \notin \{1,\ldots,h\}$. Then in the submodule $\partial_{h-1}\langle 1,\ldots,h\rangle R \cap A$ of the cyclic free bimodule $\partial_{h-1}\langle 1,\ldots,h\rangle R$, $f(\partial_{h-1}\langle 1,\ldots,h\rangle)R \in \sum_{G'} \partial_{h-1}\langle 1,\ldots,h\rangle(1-g_\alpha)R$ implies $L-G' = \emptyset$ so L is finite.

Theorem C now gives the required result.

Chen uses Theorem D plus flatness and descent of projectivity arguments to show that, if G is an Abelian group of infinite uniform dimension generated by $\aleph_m$ but no fewer elements, and if rank $G = h$, then $\mathrm{gl.d}(K_\phi[G]) = m+h+1$. He also shows that the infinite uniform dimension hypothesis can be dropped if $h \leq 1$. We conclude this note by observing that Chen's descent of projectivity results lead to a (relatively) easy proof of the global dimension of $K_\phi[G]$ in the case that $\mathrm{char}(K) = 0$.

<u>Theorem E</u> (Chen [2]). Let $f\colon R \to S$ be a ring monomorphism such that $0 \to R_R \xrightarrow{f} S_R$ is pure exact. Then if M_R is flat, M_R is R-projective iff $M \otimes_R S$ is S-projective.

The proof is just non-commutative Gruson-Raynaud (see [9]) once Chen shows that, for M_R flat, M_R is R-Mittag Leffler iff $M \otimes_R S$ is S-Mittag Leffler. It is not true that $M \otimes_R S$ flat $\Rightarrow M_R$ flat so the full impact of the commutative theory does not go over to the non-commutative case.

<u>Theorem F</u>. Let K be a field of characteristic 0, G an Abelian group of rank h generated by $\aleph_m$ but no fewer elements, $\phi\colon G \to \mathrm{Aut}(K)$. Then $\mathrm{gl.d}(K_\phi[G]) = m+h+1$.

Proof. Any basis for K over $\mathbf{Q}$ which contains 1 is a free basis for $K_\phi[G]$ over $\mathbf{Q}[G]$. Hence $0 \to \mathbf{Q}[G] \to K_\phi[G]$ is pure exact (indeed split). Moreover, for each torsion free $g \in G$, $(1-g)$ is a central non-unit, non-zero divisor of $\mathbf{Q}[G]$. Standard commutative algebra shows that, for $\{g_i \mid 1 \leq i \leq h\} \subseteq G$ generating a free subgroup of rank h, both the weak global dimension and the global dimension of $\mathbf{Q}[G]$ are obtained from the corresponding dimension of $\mathbf{Q}[G/\sum_{i=1}^{h} g_i\mathbf{Z}]$ by adding h. Thus any ideal of $\mathbf{Q}[G]$ has weak dimension $\leq h-1$ and by minor observations plus Theorem D, Case i, (see [4] for the first proof) $\mathrm{gl.d}(\mathbf{Q}[G/\sum_{i=1}^{h} g_i\mathbf{Z}]) = m$. Select any ideal I of $\mathbf{Q}[G]$ with $\mathrm{pd}(I) = m+h-1$. Let

$$\cdots \to Q_n \to \cdots \to Q_0 \to I \to 0$$

be a $\mathbf{Q}[G]$-projective resolution of I. Since ${}_{\mathbf{Q}[G]}K_\phi[G]$ is flat, tensoring gives a projective resolution of the ideal $IK_\phi[G] \approx I \otimes_{\mathbf{Q}[G]} K_\phi[G]$, and Theorem E says $\mathrm{pd}(IK_\phi[G]) = m+h-1$ so $\mathrm{gl.d}(K_\phi[G]) \geq m+h$. The reverse inequality is typical Auslander's Lemma.

BIBLIOGRAPHY

1] M. Auslander, On the dimension of modules and algebras. III: Global dimension, Nagoya Math. J. 9 (1955), 67-77.

2] N. Chen, Global dimension of skew group rings, Ph.D. thesis, Rutgers University, 1978.

3] B. Osofsky, Homological dimension and the continuum hypothesis, Trans. Amer. Math. Soc. 132 (1968), 217-230.

4] B. Osofsky, Homological dimension and cardinality, Trans. Amer. Math. Soc. 151 (1970), 641-649.

5] B. Osofsky, Hochschild dimension of a separably generated field, Proc. Amer. Math. Soc. 41 (1973), 24-30.

6] B. Osofsky, Homological dimensions of modules, CBMS Regional Conference Series in Math. 12, Amer. Math. Soc., Providence, 1973.

7] B. Osofsky, Projective dimension of "nice" directed unions, Journal of Pure and Applied Algebra, 13 (1978), 417-457.

8] R. Pierce, The global dimension of Boolean rings, J. Alg. (1967), 91-99.

9] M. Raynaud and L. Gruson, Critères de platitude et de projectivité, Invent. Math. 13 (1971), 1-89.

ACTIONS DE GROUPES ET ANNEAUX REGULIERS INJECTIFS

par

Guy RENAULT

I.- INTRODUCTION.

Soient R un anneau unitaire et G un groupe fini d'automorphismes de R . On notera R^G l'anneau des invariants de R et si R est à idéal singulier à droite nul, $\hat{R}$ (resp. $\widehat{R^G}$) désignera l'enveloppe injective de R_R (resp. $R^G_{R^G}$) . On sait que dans ce cas, G se prolonge à $\hat{R}$ (9) et l'on notera également G ce prolongement.

Dans la première partie, consacrée aux anneaux réduits (i.e. sans éléments nilpotents ≠ o) , on montre notamment que $\hat{R}$ est réduit si, et seulement si, $\hat{R}^G$ est réduit et dans ce cas, on a $\hat{R}^G = \widehat{R^G}$. Cela permettra de donner de nouvelles démonstrations de théorèmes énoncés par V. Kharchenko (7). On prouvera également que si R est un anneau réduit injectif alors R est un R^G-module projectif de type fini engendré par $n \leq |G|$ générateurs.

Dans la seconde partie, on supposera toujours que l'ordre de G est non diviseur de zéro dans R et l'on comparera les types de R et R^G dans le cas de type I (11),(5). On établira en particulier que si R est un produit fini d'anneaux de matrices sur des anneaux réduits injectifs, alors R est un R^G-module de type fini. On généralisera ce résultat lorsque R est un anneau à I.P. (avec ou sans $|G|$-torsion). Pour un exposé général sur la théorie des actions de groupes, on pourra se reporter à (9).

II.- ETUDE DES ANNEAUX REDUITS.

Pour les propriétés classiques des anneaux réduits, on pourra consulter (13,p.49).

HEOREME 1.- *Soit* G *un groupe fini d'automorphismes d'un anneau réduit* R .
s conditions suivantes sont équivalentes :

1) L'enveloppe injective de R *est un anneau réduit.*

2) L'enveloppe injective de R^G *est un anneau réduit.*

ns ces conditions, l'enveloppe injective de R^G *est l'anneau des invariants l'enveloppe injective de* R .

D'après (2) , on sait que l'on a $\hat{R} = h\hat{R} \oplus (1-h)\hat{R}$, où h est un idem-
otent central de $\hat{R}$, $(1-h)\hat{R}$ est un anneau réduit, $h\hat{R}$ est un anneau
type III. L'unicité d'une telle décomposition permet de dire que h est
nvariant par G .

) $\Longrightarrow$ 2) D'après (10) , il suffit de prouver que si x , y sont deux élé-
ents de R^G tels que $xR^G \cap yR^G = o$, alors on a $xy = o$. Supposons
$\neq o$, $y \neq o$; on a $x^2R \cap y^2R = o$ (7, p.145) . Sinon, soit $J = \{a \in R/x^2a \in y^2R\}$;
J est un idéal à droite invariant $\neq o$ de R il existe dont (7) $b = xj$
$\neq o$, élément de $xJ \cap R^G$. x^2j est un élément invariant et l'on a
$^2j = y^2c = y^2c^g$, $\forall g \in G$ et l'élément yc est un élément $\neq o$ de
$R^G \cap yR^G$, d'où la contradiction. Comme $\hat{R}$ est un anneau réduit, on a
$^2y^2 = o = xy = yx$, d'où la propriété.

) $\Longrightarrow$ 1) La démonstration résultera des lemmes suivants. Soit k un
dempotent central de $\hat{R}$ invariant par G .

EMME A.- $R[k]$ *est un anneau réduit.*

oit la relation $(r_1+r_2k)^2 = o$, avec $r_1, r_2 \in R$. On a $r_1^2+(r_1r_2+r_2r_1+r_2^2)k=o$,
e qui implique $r_1^2(1-k) = o$. Il existe un idéal à droite J essentiel
ans R tel que $(1-k)J \subset R$. D'après les propriétés classiques des anneaux
éduits, on obtient :

$$r_1^2(1-k)J = o = r_1(1-k)J .$$

Comme $\hat{R}$ est un R-module à sous-module singulier nul on a $r_1 = r_1 k$ ce qui permet de supposer que $r_1 = o$. On a alors $r_2^2 k = o$ et il suffit de reprendre la méthode précédente pour conclure.

<u>LEMME B</u>.- $R[k]^G$ *est extension essentielle de* R^G .

Soit $x+yk$ un élément $\neq o$ de $R[k]^G$. Il existe un idéal à droite essentiel et invariant de R tel que l'on ait $(x+yk)J \subset R$. On a de plus $(x+yk)J \cap J \cap R^G \neq (o)$ (7). Il existe alors une relation de la forme

$$(x+yk)j_1 = j_2 \neq o \quad , \quad j_1 \in J \quad , \quad j_2 \in J \cap R^G$$

$(x+yk)j_2$ n'est pas nul ; sinon, on aurait $(x+yk)^2 j_1 = o$, mais comme $R[k]$ est réduit (lemme A), cela impliquerait $(x+yk)j_1 = o$. On a donc $(x+yk)j_2 \in (x+yk)R^G \cap R^G - \{o\}$.

Supposons que $\hat{R}$ ne soit pas réduit ; d'après le lemme B et (2) on sait que $\widehat{R[k]}^G$ est un anneau réduit. On est donc ramené à la situation suivante : $\widehat{R^G}$ est un anneau réduit et $\hat{R}$ est un anneau de type III. Soit L un idéal à droite de R maximal pour la propriété $L \cap R^G = (o)$. Si K est un complément de L dans R , nous allons montrer que la relation $x , y \in K$ $xR \cap yR = o$ implique $xy = o$; ce qui d'après (10) impliquera que $\hat{R}$ n'est pas de type III, d'où la contradiction. D'après la maximalité de L on a :

$$(L \oplus xR) \cap R^G \neq o \quad , \quad (L \oplus yR) \cap R^G \neq o \, .$$

Soit J l'idéal à droite de R engendré par l'ensemble S .

$$S = \{a \in R / \exists\, \ell \in L \quad \text{avec} \quad \ell + xa \in R^G\}$$

et soit J' l'idéal à droite engendré par S'

$$S' = \{b \in R / \exists\, \ell' \in L \quad \text{avec} \quad \ell' + xb \in R^G\}$$

(resp. J') est un idéal à droite essentiel dans R^G . En effet, soit $\in R$. Si $xr = o$, $r \in J$, sinon, on a d'après la maximalité de L $L \oplus x r R) \cap R^G \neq o$, et R^G contient un élément $\neq o$ de la forme $\ell + x r \lambda$; omme on a $L \cap R^G = o$, on en déduit $x r \lambda \neq o$ et $r \lambda \neq o$ est un élément e J . Soit a (resp. b) un élément de S (resp. S'). Il existe donc ℓ , $' \in L$ tels que $\ell + x a \in R^G$, $\ell' + y b \in R^G$. On a $(L \oplus x R) \cap (L \oplus y R) = (o)$, ar x et y appartiennent à K , on en déduit $(\ell + x a) R^G \cap (\ell' + y b) R^G = o$. 'enveloppe injective de R^G étant réduite, il vient $(\ell + xa)(\ell' + yb) = o$ oit :

$$\ell\ell' + \ell y b + x a \ell' + x a y b = o$$

e qui implique $x a \ell' + x a y b = o$, R étant réduit on obtient

$\ell' x a + y b x a = o$ d'où $y b x a = o = x a y b$. On a alors $a y J' = o$ et comme J' est essentiel dans R , $x a y$ est nul ; $= x a y = y x a$, ce qui entraîne $y x J = o$. J étant essentiel, on a inalement $y x = o$ ce qui achève la démonstration. Montrons maintenant ue $\hat{R}^G = \widehat{R^G}$. Il est facile de montrer que $\hat{R}^G$ est un anneau injectif [9]. 'autre part, on vérifie facilement que si I est un idéal à droite essentiel ans R^G , alors IR est essentiel dans R . Le fait que $\hat{R}^G$ soit l'enveloppe njective de R^G résultera de la propriété générale suivante :

<u>EMME C.</u>- *Si* J *est un idéal à droite essentiel dans un anneau réduit* R , *lors* $J \cap R^G$ *est essentiel dans* R^G .

oit $a \neq o$ un élément de R^G , il existe un idéal à droite essentiel et nvariant T de R tel que $a T \subset J$. On a alors :

$$a T \cap T \cap R^G \neq (o)$$

n conclut alors comme dans la démonstration du lemme B .

COROLLAIRE 1. (7)- *Soit* G *un groupe fini d'automorphismes d'un anneau réduit* R . *Les conditions suivantes sont équivalentes :*

1) R *est un anneau de Goldie.*

2) R^G *est un anneau de Goldie.*

En effet, $\hat{R}$ ou $\widehat{R^G}$ est un produit fini de corps. On peut également utiliser le fait que J est essentiel dans R_R si, et seulement si $J \cap R^G$ est essentiel dans R^G .

THEOREME 3.- *Soit* G *un groupe fini d'automorphismes d'un anneau réduit auto-injectif* R . *Alors* R *est un* R^G*-module projectif de type fini engendré par* $c \leq |G|$ *générateurs.*

LEMME D.- *Soit* m *un idéal maximal du centre de* R^G. $(R/mR)^G$ *est isomorphe à* R^G/mR^G .

Soit $x \in R$, tel que $x-x^g \in mR$ pour tout $g \in G$. Il existe un idempotent central de R^G , h_g , $h_g \notin m$ tel que $h_g x = h_g x^g$. Si l'on pose $h = \prod_g h_g$, on a $x \equiv hx(mR)$ et l'assertion résulte du fait que $h\,x$ est invariant.

LEMME E. - *Si* R *est un produit fini de corps commutatifs ou non, alors* R *est un* R^G*-module engendré par* $c \leq |G|$*-générateurs.*

Soit $R = Re_1 \times \ldots \times Re_n$ la décomposition de R en produit de corps. Si H_i désigne le stabilisateur de e_i , on vérifie sans peine que l'on a $(Re_i)^{H_i} = R^G e_i$. Comme Re_i est un corps on a $Re_i : R^G e_i \leq |H_i| \leq |G|$ et l'assertion en résulte.

D'après le théorème 1, $\hat{R}$ est un anneau réduit, on en déduit facilement que R/mR est un anneau réduit et comme R^G/mR^G est un corps, R/mR est un produit fini de corps. D'après le lemme E , R/mR est un module de type fini sur R^G engendré par $s \leq |G|$ éléments. Pour tout $x \in R$, $x R^G$ est un module injectif, on peut donc écrire

$$R = \bigoplus_{i=1}^{s} x_i R^G \oplus K$$

avec
$$R_m = \bigoplus_{i=1}^{s} x_i R_m^G \quad , \quad K_m = o$$

On choisit m de telle sorte que s soit maximum. On pose $\ell_{R^G}(x_i) = h_i R^G$, $h_i \in m$, $(1-h)R^G = \sum_{i=1}^{s} h_i R^G$. Dans l'anneau $R' = R h$ on a $R' = \bigoplus_{i=1}^{s} R^G h x_i \oplus Kh$, avec $\ell_{R^G h}(x_i) = o$.

Soit $m'h$ un idéal maximal du centre de $R^G h$; on a :

$R_{m'} = \bigoplus_{i=1}^{s} x_i (K^G h)_{m'} \oplus K_{m'}$ et comme s est maximum, $K_{m'} = o$ ce qui implique $Kh = o$. On en déduit que Rh est un $R^G h$-module libre de rang $s \leq |G|$. Soit $(h_i)_{i \in I}$ une famille orthogonale maximale d'idempotents centraux de R^G tels que $R h_i$ soit libre sur $R^G h_i$ de rang $s_i \leq |G|$ et soit $h_o = \sup_i (h_i)$. Si h_o est différent de 1 , $R(1-h_o)$ contient d'après ce qui précède un idempotent k tel que Rk soit libre sur $\hat{R}^G k$, d'où la contradiction. Comme R est injectif, on a $R = \prod_i R h_i$, et R est projectif de type fini sur R^G .

<u>Remarque</u> : L'exemple suivant montre que la propriété R^G auto-injectif n'implique pas R auto-injectif.

Soit A l'anneau produit $\mathbb{C}^{\mathbb{N}}$ muni de son involution évidente, on considère le sous-anneau R des éléments $(x_i)_{i \in \mathbb{N}}$ tels que $x_i \in \mathbb{R}$ à partir d'un certain rang. A est l'enveloppe injective de R et l'on a $A^G = R^G = \mathbb{R}^{\mathbb{N}}$.

COROLLAIRE 4. (7)- *Soit* G *un groupe fini d'automorphismes d'un anneau réduit* R . *Si* R^G *vérifie une identité standard* S_n *de degré* n , *alors* R *vérifie* $S_{n|G|}$.

D'après (13, p.162) , $\widehat{R^G}$ vérifie l'identité standard S_n , c'est donc un anneau réduit. D'après le théorème 3, $\hat{R}$ se prolonge dans $M_{|G|}(\widehat{R^G})$ d'où l'assertion. On suppose que R est sans $|G|$-torsion:

PROPOSITION 5.- *Si* R *est un anneau réduit, il existe un idéal à droite* J *essentiel dans* R , *qui se plonge dans un* R^G*-module libre de rang* $\leqslant |G|$.

D'après le théorème 3 , on a $\hat{R} = \bigoplus_{i=1}^{s} \widehat{R^G}\, x_i$, avec $s \leqslant |G|$. On définit une application t $\hat{R}^G$-linéaire de $\hat{R}$ dans $(\hat{R}^G)^s$ en posant

$$t(r) = (tr(x_i r))_{1\leqslant i\leqslant s}$$

La relation $t(r) = o$ implique $t(\hat{R}r) = o$ et comme $\hat{R}$ est un anneau réduit $\hat{R}r$, est un idéal bilatère. On en déduit que ker t est un idéal bilatère invariant de $\hat{R}$. Si l'on a ker t $\neq$ o , il existe $a \neq o$, $a \in \ker t \cap R^G$. On en déduit en particulier $tr(x_i a) = o = (tr\, x_i)a$ pour tout i . Or on a une relation de la forme

$$1 = \sum_{i=1}^{s} \lambda_i\, x_i \quad , \quad \lambda_i \in \hat{R}^G$$

ce qui implique

$$|G| = \sum_{i=1}^{s} \lambda_i (tr\, x_i)$$

on en déduit alors $|G|\, a = o$ d'où la contradiction, ce qui prouve que ker t est nul. Il existe un idéal à droite J essentiel dans R tel que l'on ait $x_i\, J \subset R$ pour tout i . Pour tout $j \in J$, $tr(x_i j)$ appartient à R^G d'après ce qui précède t est une injection de J dans $(R^G)^s$.

Remarque : Le théorème 3 ne se généralise pas à un anneau réduit R , même si R est de Goldie. On sait qu'il existe un anneau commutatif intègre et noethérien R un groupe fini d'automorphismes G de R tels que R^G ne soit pas noethérien. D'après le théorème d'Eakin Nagata R n'est pas un R^G-module de type fini.

COROLLAIRE 6. (3)- *Si* R *ou* R^G *est un anneau de Goldie réduit, alors* R *se plonge dans* $(R^G)^{|G|}$.

Compte-tenu du corollaire 1, tout idéal essentiel de R contient un non-diviseur de zéro.

III.- ANNEAUX REGULIERS AUTO-INJECTIFS A DROITE DE TYPE I.

Dans ce paragraphe, sauf mention expresse du contraire, l'ordre du groupe est non diviseur de zéro dans R (R est sans |G|-torsion). Pour la théorie des anneaux de type I, on pourra consulter (5),(11).

Le théorème suivant de A. Page sera d'une grande utilité.

THEOREME 7. (9)- *Soient* R *un anneau semi-premier à idéal singulier nul,* G *un groupe fini d'automorphismes de* R . *Si* R *est sans* |G|*-torsion,* $(\hat{R})^G$ *est l'enveloppe injective de* R^G.

On rappelle que R est un anneau de type I_{fin} si, et seulement si, c'est un produit d'anneaux de matrices sur des anneaux réduits auto-injectifs, en particulier R est auto-injectif.

PROPOSITION 8.- *Soit* R *un anneau de type* I_{fin} *(resp.* I_∞*).* R^G *est alors de type* I_{fin} *(resp.* I_∞*).*

D'après (11, théorème 3.7), on peut supposer que $R = M_n(A)$ où A est un anneau réduit auto-injectif. R^G est alors un anneau régulier auto-injecti (9) qui vérifie la propriété suivante : pour tout idéal maximal P de R^G, R^G/P est un anneau simple, ce qui impliquera d'après (11, prop. 3.10) que R^G est de type I . En effet, soit m un idéal à droite maximal de R^G tel que $P = r(R^G/m)$. On a $mR \neq R$ car $|G|$ est inversible, il existe donc un idéal à droite maximal M de R tel que $M \cap R^G = m$. On sait alors (8) que R/M est un R^G-module semi-simple, par suite il existe un idéal maximal P' de R tel que $P' \cap R^G$ soit inclus dans P . R/P étant un anneau simple, c'est un R^G-module semi-simple (8) , on en déduit alors que R^G/P est de longueur finie, d'où l'assertion.

Supposons maintenant que R^G soit de type I_∞ (i.e. R ne contient que des facteurs de type I_{fin}) . On a $R^G = h\,R^G \oplus (1-h)R^G$ où $(1-h)R^G$ est de type I et où $h\,R^G$ ne contient pas d'idempotents abéliens. Comme $|G|$ est inversible dans R on a $(h\,R\,h)^G = h\,R^G\,h$. Pour montrer que $h = o$, il suffit de supposer que R^G ne contient pas d'idempotents abéliens $\neq o$. Soit m un idéal maximal du centre Z de R . D'après (11, prop.2.9) , on sait que $B = R/\cap\, m^\sigma R$, $\sigma \in G$, est un produit fini d'anneaux réguliers dont le socle est essentiel dans B . On en déduit (9) que le socle de $R^G/R^G \cap (\cap\, m^\sigma R)$ est différent de o .

<u>LEMME E.</u>- *Soient* R *un anneau régulier auto-injectif à droite et* I *un idéal bilatère de* R *tel que le socle de* R/I *soit non nul. Alors* R *contient un idempotent abélien* $\neq o$.

Soit e un idempotent de R tel que $\bar{e}\,\bar{R}$ soit simple. D'après (11, p.246) il existe des idempotents orthogonaux e_1 , e_2 , e_3 tels que

$R = e_1 R \oplus e_2 R \oplus e_3 R$ les idéaux $e_1 R$, $e_2 R$ étant isomorphes, e_3 étant un idempotent abélien. On a alors $\bar{e}\bar{R} = \bar{e}_1 \bar{A} \oplus \bar{e}_2 \bar{A} \oplus \bar{e}_3 \bar{A}$, ce qui implique que I contient e_1 et e_2 et l'on a $e_3 \neq o$.

On en déduit que R^G est de type I. Si R^G admet une composante de type I_{fin} $R^G h$, il existe un idéal maximal M de R^G contenant h tel que R^G/M soit simple. $M = \bigcap_{i=1}^{n} m_i$, où m_i est un idéal à droite maximal de R^G et l'on peut trouver des idéaux à droite maximaux de $R\, m'_i$, $1 \leqslant i \leqslant n$, tels que l'on ait $m_i = R^G \cap m'_i$. $R/\bigcap_i m'_i$ est un R^G-module semi-simple [8] et il existe des idéaux primitifs $P_1, \ldots, P_s$ de R tels que $(\bigcap_{i=1}^{s} P_i) \cap R^G \subset M$. Comme $1-h \in m_i$ pour tout i, $1-h \in m'_i$ pour tout i, et par suite $1-h \in \bigcap_{i=1}^{s} P_i \cap R^G$. On en déduit :

$$(R/\bigcap_i P_i)^G \simeq R^G h / (\bigcap_{i=1}^{s} P_i \cap R^G h)$$

Ce dernier anneau étant semi-simple, $R/\bigcap_i P_i$ est semi-simple ; d'après la proposition 3.10 de [11], R admet une composante de type I_{fin} d'où la contradiction.

<u>COROLLAIRE 9</u>.- *Soit* R *un anneau régulier auto-injectif de type* I. *Si* R^G *est de type* I_{fin} *(resp.* I_∞*), alors* R *est de type* I_{fin} *(resp.* I_∞*).*

D'après la démonstration de la proposition 8, on sait que si R^G est de type I_{fin}, alors R est de type I_{fin} ; le cas du type I_∞ s'en déduit compte-tenu de la proposition 8.

<u>COROLLAIRE 10</u>.- *Si* R *est un anneau de type* II, *alors* R^G *est de type* II.

Compte-tenu de la proposition 8, le corollaire résulte facilement de la propriété suivante : soit e un idempotent fini de R ; on pose $fR = \sum_{\sigma \in G} e^\sigma R$, $f \in R^G$. D'après [11, prop. 5.1] f est un idempotent fini de R, donc de R^G.

THEOREME 11.- *Soit* G *un groupe fini d'automorphismes d'un anneau* $R = \prod_{i=1}^{s} M_{n_i}(A_i)$ *où* A_i *est un anneau réduit auto-injectif. Alors* R *est un* R^G*-module projectif de type fini, engendré par* $c \leq |G|(\sup_i n_i)$ *éléments.*

D'après (11, p.245) il suffit de considérer le cas où $R = M_n(A)$, A réduit auto-injectif. $Z(R)$ (resp. $Z(R)^G$) désignera le centre de R (resp. R^G) . $Z(R)$ est entier sur $Z(R)^G$ et si m est un idéal maximal de $Z(R)^G$, on vérifie facilement que R/mR (resp. $R^G/m\,R^G$) est un anneau semi-simple. Comme on a $m \subset R^G$, $m\,R \cap R^G = m\,R^G$ ce qui implique

$$(R/m\,R)^G \simeq R^G/m\,R^G$$

D'après (8) . $R/m\,R$ est un R^G-module de longueur finie engendrée par $c \leq |G|\ \ell(R/m\,R)$ éléments, où $\ell(.)$ désigne la longueur. $Z(R)/m\,Z(R)$ est un produit de $s \leq |G|$ corps par suite $m\,R$ est l'intersection de $s \leq |G|$ idéaux maximaux de R . Ce qui implique $\ell(R/m\,R) \leq |G|n$. On choisit m tel que c soit maximum. Pour tout $x \in R$, $x\,R^G$ est un R^G-module injectif on peut donc écrire :

$$R = x_1\,R^G \oplus \ldots \oplus x_c\,R^G \oplus K \quad , \quad \text{avec} \quad K = m\,K \ .$$

Il suffit de reprendre la démonstration du théorème 3 pour conclure.

PROPOSITION 12.- *Soit* R *un anneau semi-premier à idéal singulier nul tel que son enveloppe injective soit produit d'un nombre fini d'anneaux de matrices sur des anneaux réduits. Il existe alors un idéal à droite essentiel* J *de* R *qui se plonge dans un* R^G*-module libre de rang fini.*

La démonstration est analogue à celle de la proposition 5 et l'on utilise de plus le fait que $|G|$ est inversible dans $\hat{R}$.

COROLLAIRE 13. [3]- *Si* R *ou* R^G *est un anneau de Goldie, alors* R *se plonge dans un* R^G*-module libre.*

COROLLAIRE 14. [7]- *Si* R *est un anneau semi-premier tel que* R^G *soit à I.P. Alors* R *est un anneau à I.P.*

C'est une conséquence des théorèmes 7 et 11, compte -tenu du fait que $\hat{R}^G$ est à I.P. [19, p.162].

Lorsque l'anneau régulier auto-injectif R admet de la $|G|$-torsion on a un résultat analogue au théorème 11, mais en supposant que l'anneau R est à identité polynomiale. Nous allons tout d'abord préciser un résultat de [1].

PROPOSITION 15.- *Soit* R *un anneau régulier auto-injectif vérifiant une identité polynomiale de degré* d *. Alors* R *est un* Z(R)*-module projectif de type fini, engendré par* $c \leq \left[\frac{d}{2}\right]^2$ *éléments.*

Soit m un idéal maximal du centre ; comme R est un Z(R)-module du type fini [1] , le centre de R/m R est Z(R)/m [12] , d'après le théorème de Kaplansky, R/mR admet un système de $s \leq \left[\frac{d}{2}\right]^2$ générateurs. On choisit m de telle sorte que s soit maximum. Il suffit alors de recopier la fin de la démonstration du théorème 3 pour conclure.

THEOREME 16.- *Soit* G *un groupe fini d'automorphismes d'un anneau régulier* R *auto-injectif vérifiant une identité polynomiale de degré* d *(*R *avec une* $|G|$*-torsion). Alors* R *est un* R^G*-module (resp.* $Z(R)^G$*-module) engendré par* $n \leq \left[\frac{d}{2}\right]^2 |G|$ *éléments.*

D'après la proposition 15, R est un Z(R)-module de type fini engendré par $c \leq \left[\frac{d}{2}\right]^2$ éléments et comme Z(R) est un anneau réduit auto-injectif, il suffit d'appliquer le théorème 3 pour conclure.

BIBLIOGRAPHIE

(1) E.P. Armendariz and S.A. Steinberg. Regular self-injective rings with a polynomial identity (à paraître)

(2) A. Cailleau-Page & G. Renault. Sur l'enveloppe injective des anneaux semi-premiers à idéal singulier nul. J. of Algebra,45(1970)133-141.

(3) D.R. Farkas & R.L. Snider. Noetherian fixed rings. Pacific J. of Math., 69(1977)347-353.

(4) J.W. Fisher et J. Osterburg. Semi-prime ideals in rings with finite group actions. J. of Algebra, 50(1978)488-502.

(5) K.R. Goodearl and A.K. Boyle. Dimension theory for non singular injective modules. Memoirs Amer. Math. Soc. n° 177 (1976).

(6) V.K. Kharchenko. Galois extensions and quotient rings. Algebra and Logic, Nov. 1975, 265-281.

(7) V.K. Kharchenko. Generalized identities with automorphism. Algebra and Logic, Mars 1976, 132-148.

(8) M. Lorenz. Primitive ideals in crossed products and rings with finite group actions (à paraître).

(9) A. Page. Actions de groupes. Séminaire P. Dubreil 1977-78. Lecture Notes in Mathematics, Springer Verlag (à paraître)

(10) G. Renault. Anneaux réduits non commutatifs. J. Math. Pures et Appl. 4 (1967).

(11) G. Renault. Anneaux réguliers auto-injectifs à droite. Bull. Soc. Math. France, 101 (1973).

(12) G. Renault. Anneaux biréguliers auto-injectifs à droite. J. algebra vol. 36, n°1 (1975)

(13) G. Renault. Algèbre non commutative. Gauthier-Villars, Paris (1975)

Université de Poitiers
40, avenue du Recteur Pineau
86022 POITIERS (FRANCE)

Il faut remarquer que certains résultats présentés ici sont améliorés dans l'article, Actions de groupes, par D. Handelman et G. Renault.

K_2 of some truncated polynomial rings

by

Leslie G. Roberts[2]

(section 7 written jointly with S. Geller[1]).

1. Introduction. Let k be a commutative ring. In this paper I study K_2 of $k[X]/X^n$, $k[X,Y]/(X,Y)^n$ and similar "truncated polynomial rings". My original motivation was the study of K_1 of lines through the origin in the plane [10]. In [10] k had to be a field. In §4 of this paper I use the presentation for K_2 given in [8] to do the same computation for k a more general commutative ring. More recently, when studying excision for K_1, I have needed to know $K_2k[X]/X^n$ for various rings k. Theorems 3 and 4 of this paper will be used in a revised version of [5].

I have known most of the results in §2-6 for two years. Some of these results have been noticed independently by others. I am now aware of the following: (1) [2(b)] compares the homomorphism ω of §3 with the homomorphism $K_2(R) \to \Lambda^2\Omega_R$ defined by Gersten in [5(b)] p 238,

---

[1]Partially supported by NSF-MCS77-00096.

[2]Partially supported by the National Research Council of Canada Grant A7209.

(2) [13] contains some (but not all) of the results in §2, 3, 5. (3) The field case of Theorem 3 is included in lemma 3.4 of [3]. (4) Stienstra's thesis contains extensive computations in characteristic p, but I have not seen it yet.

Just before this conference I managed to compute $K_2Z[X]/X^3$ and $K_2Z[X]/X^4$ directly from the presentation of [8]. This work is written up in [10(a)]. At the conference Keith Dennis outlined to Sue Geller and me the approach used in §7, and kindly permitted us to use his ideas.

For the convenience of the reader I will recall the presentation of [8]. Let R be an augmented k-algebra with nilpotent augmentation ideal I. Then $K_2(R,I) = D(R,I) = \tilde{K}_2(R) = \ker[K_2(R) \to K_2(k)]$ is the abelian group with generators $\langle a,b \rangle$ such that $a \in I$ or $b \in I$, and relations

$$(D1) \quad \langle a,b\rangle\langle -b,-a\rangle = 1$$

$$(D2) \quad \langle a,b\rangle\langle a,c\rangle = \langle a,b+c+abc\rangle$$

$$(D3) \quad \langle a,bc\rangle = \langle ab,c\rangle\langle ac,b\rangle$$

It is proved in [8] that in D3 it suffices to consider the case where one of a, b, or c lies in I. From D3 it follows that $\langle a,1\rangle = 1$ for all $a \in I$. If we use D1 to take the inverse of both sides in D2, we get

$<-b,-a><-c,-a> = <-b-c-abc,-a>$ and replacing a, b, c by -a, -b, -c respectively, we get

$$(D2)' \quad <b,a><c,a> = <b+c+abc,a>$$

Repeated application of D3 proves that $<a,b^n> = <ab^{n-1},b>^n$.

If sufficiently many integers are invertible in k then $K_2(R,I)$ can be described in terms of differentials. This result is due to Bloch [2(a)]. After the presentation of [8] became available this result reappeared as example 3.12 in [8] which I describe in §2 and then use to put into more explicit form K_2 of some of the truncated polynomial rings that I am interested in.

2. Bloch's Theorem.

Let R be an augmented k-algebra with augmentation ideal J . Assume that $J^N = 0$ for some $N \geq 1$, and that $N!$ is a unit in k . Then Bloch's theorem says that $K_2(R,J) \cong \Omega^1_{R,J}/dJ$. For $K_2(R,J) = D(R,J)$ we use the presentation in terms of pointy brackets < > and relations D1, D2, D3 given in [8]. By definition $\Omega^1_{R,J} = \ker[\Omega^1_{R/Z} \to \Omega^1_{k/Z}]$ (the homomorphism being induced by the augmentation homomorphism $R \to k$). First I give a presentation for $\Omega^1_{R,J}/dJ$ and then describe the isomorphism $K_2(R,J) \to \Omega^1_{R,J}/dJ$. ($\Omega^1_{R/Z}$ is the absolute Kahler differential of R. Later the Z and the superscript 1 will be omitted).

First note that Ω^1_R is the abelian group with generators $adb (a,b \in R)$ and relations $ad(b_1+b_2) = adb_1 + adb_2$, $(a_1+a_2)db = a_1db + a_2db$, and $ad(b_1b_2) = ab_1db_2 + ab_2db_1$. This is an R-module via $r\cdot(adb) = (ra)db$. The split surjection $R \to k$ yields a direct sum decomposition $\Omega^1_R = \Omega^1_{R,J} \oplus \Omega^1_k$ so $\Omega^1_{R,J} \cong \Omega^1_R/$(subgroup generated by $\alpha d\beta$, $\alpha,\beta \in k$). If $a \in R$ denote its image in k by $\bar{a}$. Then the projection of adb onto $\Omega^1_{R,J}$ is

$$adb - \bar{a}d\bar{b} = adb - \bar{a}db + \bar{a}db - \bar{a}d\bar{b} = (a-\bar{a})db + \bar{a}d(b-\bar{b}).$$

Therefore $\Omega^1_{R,J}$ is the subgroup of Ω^1_R generated by all adb in which either a or b lies in J. I now claim that $\Omega^1_{R,J}$ is the abelian group with generators adb ($a,b \in R$, either a or $b \in J$) and relations

($\Omega 2$) $ad(b+c) = adb + adc$ ($a \in J$ or b and $c \in J$)

($\Omega 3$) $ad(bc) = abdc + acdb$ (a,b, or $c \in J$)

($\Omega 4$) $(a+b)dc = adc + bdc$ ($c \in J$ or $a,b \in J$)

The numbering is the same as in [8]. Relation $\Omega 4$ is left out of [8], but it can be derived from Ω_2 and Ω_3 by applying Ω_2 and Ω_3 to $d((a+b)c)$. I feel more comfortable including Ω_4. Let K be the group with this presentation. Then there is a surjection $K \to \Omega^1_R/$ {subgroup generated by $\alpha d\beta$, $\alpha,\beta \in k$} $\cong \Omega^1_{R,J}$. Simply send $adb \to adb$. Define an inverse by $adb \to \bar{a}d(b-\bar{b}) + (a-\bar{a})db$. It is then straightforward to check that this map respects the defining relations of Ω^1_R and kills the subgroup generated by $\alpha d\beta (\alpha,\beta \in k)$. Thus we have a homomorphism $\Omega^1_{R,J} \to K$ and the two homomorphisms are inverses of each other. Therefore $\Omega^1_{R,J} \cong K$.

To get a presentation of $\Omega^1_{R,J}/dJ$ we simply add the relation

$$(\Omega 1)\ 1\cdot da = 0 \quad \text{for all} \quad a \in J .$$

Note that $\Omega^1_{R,J}$ is an R-module, but dJ is only a subgroup, not an R-submodule.

The isomorphism $\sigma : K_2(R,J) \to \Omega^1_{R,J}/dJ$ is given by $\sigma\langle a,b\rangle = \ell(a,b)da$ where $\ell(a,b) =$ $b - \frac{ab^2}{2} + \frac{a^2b^3}{3} - \dots \pm \frac{a^{N-1}b^N}{N}$. This is the formal expansion of $a^{-1}\log(1+ab)$. Since a or $b \in J$ and $J^N = 0$ further terms are not needed. In fact, the final term $\frac{a^{N-1}b^N}{N}\,da$ vanishes. If $b \in J$ this is clear. If $a \in J$

then it equals $d(\frac{a^N b^N}{N^2})$ and therefore also vanishes. The inverse isomorphism is given by $\tau(adb) = \langle e(a,b),a\rangle$ where $e(a,b) = b + \frac{ab^2}{2!} + \frac{a^2b^3}{3!} + \ldots + \frac{a^{N-1}b^N}{N!}$. This is the formal expansion of $a^{-1}(\exp(ab)-1)$.

By a truncated polynomial ring I mean a ring of the form $A = k[X_1,\ldots,X_n]/(f_1,\ldots,f_m)$ where the f_i are monomials (with coefficient 1) and $f_i \nmid f_j$ for $i \neq j$. Let $J = (X_1,\ldots,X_n)A$. We assume that $J^N = 0$, for some N with $N!$ a unit of k. We want to know $\Omega^1_{A/Z}$. By Theorem 57 page 186 of [7] there is an exact sequence

$$\Omega_k \otimes_k A \xrightarrow{v} \Omega_{A/Z} \to \Omega_{A/k} \to 0$$

Let $D : A \to \Omega_k \otimes_k A$ be defined by applying $d : k \to \Omega_k$ to each coefficient. (Throughout we write each element of A in the canonical way as a polynomial in $k[X_1,\ldots,X_n]$ with no monomials in $(f_1,\ldots,f_m)k[X_1,\ldots,X_n]$ included). Then a splitting for v is given by

$$fdg \to fDg .$$

Thus $\Omega_{A/Z} \cong \Omega_{A/k} \oplus (\Omega_k \otimes_k A)$ and under this isomorphism df corresponds to $\Sigma \frac{\partial f}{\partial X_i} dX_i + Df$. Under the homomorphism $\Omega_{A/Z} \to \Omega_{k/Z}$ the term $\Omega_{A/k}$ is sent to zero because it is generated by the dX_i. Thus

$$\Omega_{A,J} \cong \Omega_{A/k} \oplus (\Omega_k \otimes_k J) .$$

Now A is a free k-module, say of rank d. Then J is free of rank d-1. $\Omega_{A/k}$ is the A-module with generators $dX_1, dX_2, \ldots, dX_n$ and relations $df_i = 0$. If $X_1^{a_1} X_2^{a_2} \ldots X_n^{a_n}$ is a monomial in J, then as an abelian group J is generated by $\lambda X_1^{a_1} X_2^{a_2} \ldots X_n^{a_n}$ $(\lambda \in k)$, so in considering dJ we need consider $d(\lambda X_1^{a_1} \ldots X_n^{a_n}) =$

$(d\lambda) X_1^{a_1} X_2^{a_2} \ldots X_n^{a_n} + \sum_{i=1}^{n} a_i \lambda X_1^{a_1} \ldots X_i^{a_i - 1} \ldots X_n^{a_n} dX_i$. Since a_i is a unit this simply eliminates one copy of k from $\oplus AdX_i$. Suppose in addition there are M monic monomials which are themselves zero in A, but which have at least one non-zero partial derivative $(M \geq m)$. If f is such a monomial then $d(0) = d(\lambda f)$ eliminates another copy of k (the latter relations correspond to factoring out by the A-submodule generated by $df_1, \ldots, df_n$ in the definition of $\Omega_{A/k}$). The free A-module on the dX_i has rank nd so altogether we are left with

$$K_2(A,J) = \Omega_{A,J/dJ} \cong (\Omega_k \otimes_k J) \oplus (nd-(d-1)-M)k .$$

(Note that each form $\lambda X_1^{a_1} X_2^{a_2} \ldots X_n^{a_n} dX_j$ can occur in only one relation - $d(\lambda X_1^{a_1} X_2^{a_2} \ldots X_j^{a_j+1} \ldots X_n^{a_n}) = 0$ so the various eliminations are independent of each other). For sake of definiteness we might eliminate those terms

$\lambda X_1^{a_1} X_2^{a_2} \ldots X_j^{a_j-1} \ldots X_n^{a_n} dX_j$ such that j is the largest index with $a_j > 0$ and $X_1^{a_1} X_2^{a_2} \ldots X_j^{a_j-1} \ldots X_n^{a_n}$ not already zero in A for example $\lambda X_1^2 dX_2$ in $k[X_1,X_2]/(X_1,X_2)^3$ or $\lambda X_1 X_2 dX_1$ in $k[X_1,X_2]/(X_1^2,X_2^2)$.

For example if $A = k[X_1,\ldots X_n]/(X_1,\ldots X_n)^N$ then d = no. of homogeneous polynomials of degree $N-1$ in $n+1$ variables $= \binom{N+n-1}{N-1} = \binom{N+n-1}{n}$, M = no. of homogeneous polynomials of degree N in n variables $= \binom{N+n-1}{N} = \binom{N+n-1}{n-1}$. For $n=2$ the expression simplifies to

$$\left(\frac{(N+2)(N-1)}{2}\, \Omega^1_{k/Z}\right) \oplus \frac{N(N-1)}{2}\, k .$$ If $A = \dfrac{k[X_1,X_2,\ldots X_n]}{(X_1^{a_1},X_2^{a_2},\ldots,X_n^{a_n})}$

then $d = a_1 a_2 \ldots a_n$ and $M = \sum_i a_1 a_2 \ldots \hat{a}_i \ldots a_n$. For $n = 2$ $K_2(A,J)$ simplifies to $(a_1 a_2 - 1)\Omega^1_{k/Z} \oplus (a_1-1)(a_2-1)k$.

. The dlog map. Consider the homomorphism $d : \Omega_R \to \Lambda^2\Omega_R$ given by $d(fdg) = df \wedge dg$. (Ω_R is an -module, and Λ^2 is taken as R-modules, rather than as -modules). Clearly dJ vanishes, so we may consider the omposition

$$K_2(R,J) \overset{\sigma}{\to} \Omega_{R,J}/dJ \overset{d}{\to} \Lambda^2\Omega_R .$$

$$\langle a,b\rangle \to \ell(a,b)da \to d\ell(a,b)\wedge da$$

ut $\ell(a,b) = b - \frac{ab^2}{2} + \frac{a^2b^3}{3} - \dots \pm \frac{a^{N-1}b^N}{N}$. so

$\ell(a,b) = (1-ab+a^2b^2 \dots \pm a^{N-1}b^{N-1})db + (*)da$.

efine φ by $\varphi\langle a,b\rangle = -d\sigma\langle a,b\rangle = \frac{da\wedge db}{1+ab}$.

Even if $N!$ is not a unit, it is straightforward o check that φ respects the defining relations D1, D2, D3 f [8], so there is a homomorphism $\varphi : K_2(R,J) \to \Lambda^2\Omega_R$ for ny augmented k-algebra R . This I will use later to study $K_2(R,J)$ when $N!$ is not a unit (in which case σ is not lefined).

I will now prove in the case of a truncated polynomial ing $A = k[X_1,\dots,X_n]/(f_1,\dots f_m)$ that $d : \Omega_{A,J}/dJ \to \Lambda^2\Omega_A$ is an inclusion. In the last section we saw that

$$\Omega_{A/Z} = \Omega_{A/k} \oplus (\Omega_k \otimes_k A) \quad \text{and} \quad \Omega_{A,J} = \Omega_{A/k} \oplus (\Omega_k \otimes_k J) .$$

Hence $\Lambda^2\Omega_A = (\Lambda^2\Omega_{A/k}) \oplus (\Omega_{A/k}\otimes_A(\Omega_k\otimes_k A)) \oplus \Lambda^2(\Omega_k\otimes_k A)$.
I will omit the third term of this direct sum decomposition since it is not needed. Restricted to a map from $\Omega_k\otimes_k J$ to $\Omega_{A/k}\otimes_A(\Omega_k\otimes_k A) = \Omega_k\otimes_k\Omega_{A/k}$, d turns out to be $1\otimes d'$ where $d'f = \sum_{i=1}^{n} \frac{\partial f}{\partial X_i} dX_i$. But d' is a split inclusion of free k-modules (here we need that $(N-1)!$ is a unit, and also that we are applying d only to J and not A). Therefore $d : \Omega_k\otimes_k J \to \Omega_k\otimes_k\Omega_{A/k}$ is an inclusion. Now let $\alpha \in \Omega_{A,J}$ with $d\alpha = 0$. Suppose $\alpha = \sum_{i=1}^{n} g_i dX_i + \sum_{i=1}^{m} \omega \otimes f_\omega$. Then $d\alpha = \sum_{i,j} \frac{\partial g_i}{\partial X_j} dX_j \wedge dX_i$ + (terms in $\Omega_k\otimes_k\Omega_{A/k}$ and $\Lambda^2(\Omega_k\otimes_k A)$). We must have $\sum_{i,j} \frac{\partial g_i}{\partial x_j} dX_j \wedge dX_i = 0$. Partial differentiation lowers the degrees of the $\frac{\partial g_i}{\partial X_j}$ below where relations amongst the $dX_j \wedge dX_i$ occur so we must have $\frac{\partial g_i}{\partial X_j} = \frac{\partial g_j}{\partial X_i}$. Thus there exists $G \in J$ so that $\frac{\partial G}{\partial X_i} = g_i$. Then $\alpha - dG \in \Omega_k\otimes_k J$ and $d(\alpha - dG) = 0$. By the above we have $\alpha = dG$. Thus the map $d : \Omega_{A,J/dJ} \to \Lambda^2\Omega_A$ is an inclusion, as required. The map d is not onto since we have ignored the term $\Lambda^2(\Omega_k\otimes_k A)$, which might be non-zero.

Thus the homomorphism φ can be used instead of σ to compute $K_2(A,J)$ when $N!$ is a unit. In fact I had computed K_2 of $k[X,Y]/(X,Y)^N$ and several other truncated polynomial rings using φ before I learned about the isomorphism σ.

4. Application to lines through the origin. Let $R = k[X,Y]/XY(Y-X)(Y-\alpha_4 X)\dots(Y-\alpha_n X)$ where $\alpha_i - \alpha_j$ are units for $i \neq j$, $i,j \geq 2$ $(\alpha_2=0, \alpha_3=1)$. The last assumption guarantees that the subscheme generated by any two of the lines is isomorphic to $k[X,Y]/(XY)$. Assume also that the homomorphisms $K_1(k) \to K_1 k[t]$ and $K_2(k) \to K_2 k[t]$ are isomorphisms.

Let $f = XY(Y-X)\dots(Y-\alpha_n X) = f_1 f_2 \dots f_n$, and $F_i = \Pi_{j \neq i} f_j$. Then $k[X,Y]/f_i \cong k[t_i]$ where $X \to 0$, $Y \to t_1$ (for i=1) and $X \to t_i$, $Y \to \alpha_i t_i$ $(i \geq 2)$. I claim that these maps induce an inclusion $R \to B = \Pi_{i=1}^{n} k[t_i]$. To show this it suffices to prove that $(\cap f_i) = \Pi f_i$ (as $k[X,Y]$-ideals). The inclusion $\supset$ is clear. If $F \in (\cap f_i)$ then $F = f_1 g_1$. But f_1 maps to a non-zero divisor in $k[t_2]$ so $g_1 \in f_2 k[X,Y]$ and $F = f_1 f_2 g_2$. Similarly f_i maps to a non-zero divisor in $k[t_j]$ $(j \neq i)$ so we can continue, yielding $F \in \Pi f_i$ as required. Now I claim that $(X^{n-1}, X^{n-2}Y, \dots, XY^{n-2}, Y^{n-1}) = (F_1, F_2, \dots, F_n)$ as R-ideals. The inclusion $\supset$ is clear. The reverse inclusion follows by mapping into B. The ideal $(F_1, \dots, F_n)$ maps to the R-module generated by $(t_1^{n-1}, 0, \dots, 0)$, $(0, t_2^{n-1}, 0, \dots, 0)$ ... (here we use that the $\alpha_i - \alpha_j$ are units), and $(X,Y)^{n-1}$ clearly maps into this. Let $I = (F_1, \dots, F_n)$. From the above we see that I is a B-ideal. Therefore we have a Cartesian square

$$\begin{array}{ccccc} & & R & \to & B = \prod_{i=1}^{n} k[t_i] \\ & & \downarrow & & \downarrow \\ k[X,Y]/(X,Y)^{n-1} & = & R/_I & \to & B/_I = \prod_{i=1}^{n} k[t_i]/(t_i^{n-1}) \end{array} .$$

Let S be any of the k-algebras R, $k[t_i]$, $k[X,Y]/(X,Y)^{n-1}$ or $k[t_i]/(t_i^{n-1})$. Then S is augmented (by evaluation at 0). Let $\tilde{K}_i(S)$ = cokernel $(K_i(k) \to K_i(S))$. The augmentation splits this so we have direct sum decompositions $K_i(S) = K_i(k) \oplus \tilde{K}_i(S)$ and $\tilde{K}_i(S)$ = kernel $(K_i(S) \to K_i(k))$ also. The ring R maps onto each factor in B so by [4] p. 246 3(c) there is a Mayer-Victoris sequence

$$\tilde{K}_2(k[X,Y]/(X,Y)^{n-1}) \xrightarrow{\lambda} \prod_{i=1}^{n} \tilde{K}_2 k[t_i]/(t_i^{n-1}) \to \tilde{K}_1(R) \to$$

$$\tilde{K}_1(R/_I) \to \prod_{i=1}^{n} \tilde{K}_1 k[t_i]/(t_i^{n-1})$$

(using the fact that $\tilde{K}_1 k[t_i] = 0$). But $K_1(R/_I) = K_1(k) \oplus K_1(R/_I, J)$ (J = augmentation ideal) so $\tilde{K}_1(R/_I) = K_1(R/_I, J) = U(R/I,J)$ $(=1+J)$ by Bass [1] p. 449. Now $U(R/I,J)$ is clearly mapped injectively into $U(B/I, \prod t_i)$ so $\tilde{K}_1(R)$ = coker λ .

Now assume that $(N-1)!$ is a unit in k, so that the $\tilde{K}_2$'s are given by Bloch's theorem. $\tilde{K}_2 k[X,Y]/(X,Y)^{n-1} = (\Omega_k \otimes_k J) \oplus k[X,Y]/(X,Y)^{n-2} YdX$ and $\tilde{K}_2 k[t]/t^{n-1} = \Omega_k \otimes_k J$. In each case J is the appropriate augmentation ideal, and he same notation is used for an element of $\Omega_{S,J}$ as its class in $\Omega_{S,J}/dJ$. Terms in $f(X)dX$, $f(X,Y)dY$, or $(t)dt$ are eliminated by dJ as indicated at the end of 2. For convenience I will write $t = (t_1,t_2,\dots,t_n)$. Then

$$\text{(a)}\quad \lambda(\omega\otimes X^iY^j) = (0^i,0^j,1,\alpha_4^j,\dots,\alpha_n^j)t^{i+j}\omega$$

$$(\omega\in\Omega_k, n-2\geq i+j\geq 1, 0^0=1)$$

$$\text{and (b)}\quad \lambda(cX^iY^jdX) = (0,0,1,\alpha_4^j,\dots,\alpha_n^j)ct^{i+j}dt$$

$$(c\in k, j\geq 1, i+j\leq n-2)$$

The terms of the second type must be expressed in terms of $\Omega_k\otimes_k J$ using the relation $dJ = 0$. Thus

$$d\,\frac{1}{i+j+1}(0,0,ct_3^{i+j+1},\alpha_4^j ct_4^{i+j+1},\dots,\alpha_n^j ct_n^{i+j+1})$$

$$= (0,0,1,\alpha_4^j,\dots,\alpha_n^j)ct^{i+j}dt + \frac{1}{i+j+1}(0,0,1,\alpha_4^j,\dots\ \alpha_n^j)t^{i+j+1}dc$$

$$+ \frac{j}{i+j+1}(0,0,0,\alpha_4^{j-1}d\alpha_4,\dots,\alpha_n^{j-1}d\alpha_n)ct^{i+j+1} = 0$$

($j\geq 1, i+j\leq n-3$ - the relations with $i+j = n-2$ vanish).

The second term on the right is contained within (a) so we can ignore it. Then $\tilde{K}_2k[t]/t^{n-1} = \bigoplus_{i=1}^{n-2}\Omega_k t^i$. The generators of image λ are homogeneous in t . For $t^i (1\leq i\leq n-2)$ we have $n\Omega_k$ factored out by the subgroup generated by

$$(0^{i-j}, 0^j, 1, \alpha_4^j, \ldots, \alpha_n^j)\omega \qquad \begin{matrix}(0 \le j \le i) \\ \omega \in \Omega_k\end{matrix}$$

and $(0,0,0,\alpha_4^{j-1} d\alpha_4, \ldots, \alpha_n^{j-1} d\alpha_n)c \quad (c \in k, 1 \le j \le i-1)$.

The cases $j=0$ and $j=i$ of the first type of relation merely eliminate the first two copies of Ω_k so we are left with $(n-2)\Omega_k$ factored out by the subgroup generated by

$$(1, \alpha_4^j, \ldots, \alpha_n^j)\omega$$

and $$(0, \alpha_4^{j-1} d\alpha_4, \ldots, \alpha_n^{j-1} d\alpha_n)c$$

$$(\omega \in \Omega_k, c \in k, 1 \le i \le n-2, 1 \le j \le i-1).$$

This is the same as expressed by the matrix M on page 357 of [10], except that here we have written $\widetilde{K}_2 k[t]/(t^{n-1}) = \bigoplus_{i=1}^{n-2} \Omega_k t^i$ instead of $\bigoplus_{i=0}^{n-3} \Omega_k t^i$ as in [10], with a corresponding shift up by 1 in the power of t. Thus as in [10] we get $\widetilde{K}_1(R) = \binom{n-1}{2}\Omega_k/V$ where V is a finitely generated k-module. If $n=3$ or if the $d\alpha_j$ are all zero, then $V = 0$. I have not checked to see if V is free, but for k a domain with quotient field K the estimates of rank V in [10] still apply because $\Omega_K = \Omega_k \otimes_k K$.

If $\widetilde{K}_2 k[t]/(t^{n-1}) = 0$ (eg k a perfect field of characteristic p, as we will see in §6) then $\widetilde{K}_1(R) = 0$.

. The case $R = k[X]/X^n$. In this section I begin xamining $\tilde{K}_2(R)$ when $n!$ is not a unit. The approach is y induction on n . The homomorphism $\tilde{K}_2(k[X]/X^n) =$ $(k[X]/X^n, Xk[X]/X^n) \to D(k[X]/X^{n-1}, Xk[X]/X^{n-1})$ is clearly nto. Denote its kernel by K . Then K is generated by ymbols $\langle a,b\rangle$ where a or b lies in $X^{n-1}k[X]/X^n$. o see this let $\alpha \in K$. Choose a representative $\hat{\alpha}$ f α in the free abelian group on symbols $\langle f,g\rangle$, $,g \in R$, f or $g \in XR$. Then the image of $\hat{\alpha}$ in the free elian group on symbols $\langle\ ,\ \rangle$ over $K[X]/X^{n-1}$ is the um of relations D1, D2, D3 . Each relation can be ifted to one over $k[X]/X^n$. Thus we may assume that $\hat{\alpha}$ as image 0 at the free abelian group level. Then the ymbols occuring in $\hat{\alpha}$ must occur in pairs $\langle f,g\rangle$; $f_1,g_1\rangle^{-1}$ with $f \equiv f_1$ and $g \equiv g_1 \bmod X^{n-1}$. But it ollows easily from D2 that $\langle f+rX^{n-1}, g+sX^{n-1}\rangle =$ $f,g\rangle\langle rX^{n-1},g\rangle\langle f,sX^{n-1}\rangle$, $r,s \in k$ (use the fact that not both and g have non-zero constant term). Using D1, D2, D3 e generators of K can be further reduced to $\langle aX^{n-1},b\rangle$ nd $\langle aX^{n-1},bX\rangle = \langle abX^{n-1},X\rangle$ (for example $\langle aX^{n-1},X^2\rangle = 1$ y D3).

By using the relations D_i it follows easily that $\to \langle aX^{n-1},X\rangle$ gives a homomorphism $\rho_2 : k \to K$. Furthermore $aX^{n-1},X\rangle^{-1} = \langle -X,-aX^{n-1}\rangle = \langle X,aX^{n-1}\rangle = \langle aX,X^{n-1}\rangle = \langle aX^{n-1},X\rangle^{n-1}$ the last step follows from repeated application of D3). hus $0 = \langle aX^{n-1},X\rangle^n = \langle naX^{n-1},X\rangle$, so we have a homomorphism

$k/nk \to K$. If $n=2$, as in [6] we get an exact sequence

$$k/2k \to K \to \Omega_k \to 0 .$$

If $n > 2$, then $adb \to \langle aX^{n-1},b\rangle$ gives a homomorphism $\rho_1 : \Omega_k \to K$. We have seen that the images of k/nk and Ω_k generate K, so there is an exact sequence

$$k/nk \oplus \Omega_k \to D(k[X]/X^n, Xk[X]/X^n) \to D(k[X]/X^{n-1}, Xk[X]/X^{n-1}) \to 0$$

$$\parallel$$

$$\tilde{K}_2(k[X]/X^n)$$

For $R = k[X]/(X^n)$ we have $\Omega_{R/k} \cong k[X]/(X^n, nX^{n-1})$ (with generator dX). Thus $\Omega_k \otimes_k \Omega_{R/k} = \bigoplus_{i=0}^{n-2} \Omega_k X^i \oplus \Omega_k/n\Omega_k X^{n-1}$ Now consider φ followed by projection onto $\Omega_k X^{n-2}dX$. We have $\langle aX^{n-1},b\rangle \to \dfrac{d(aX^{n-1})\wedge db}{1+abX^{n-1}}$ $= X^{n-1}da\wedge db + (n-1)X^{n-2}adX\wedge db \to (n-1)adbX^{n-2}dX$. Thus if $n-1$ is a unit the map of Ω_k into K is a split inclusion. Similarly $\langle aX^{n-1},X\rangle$ maps to $da \in \Omega_k/n\Omega_k$ under φ (projection not needed here as other components of $\varphi\langle aX^{n-1},X\rangle \in \Lambda^2\Omega_R$ are zero).

Our results can be summarized as :

Theorem 1 (a) Let k be any commutative ring. Then there is an exact sequence $(n>2)$.

$$k/nk \oplus \Omega_k \xrightarrow{\rho=(\rho_2,\rho_1)} \tilde{K}_2(k[X]/X^n) \to \tilde{K}_2(k[X]/X^{n-1}) \to 0$$

$$a \oplus bdc \to \langle aX^{n-1},X\rangle\langle bX^{n-1},c\rangle$$

(b) If $n-1$ is a unit in k, then ρ restricted to Ω_k is a split inclusion.

I now prove the following lemma, which will be needed later.

Lemma 2 (a) If n is a positive integer, then in $D(R,I)$, $\langle f,g\rangle^n = \langle f, \frac{(1+fg)^n-1}{f}\rangle = \langle\frac{(1+fg)^n-1}{g}, g\rangle$

(b) If fg is nilpotent the above formulas also hold for n a negative integer.

Proof. First assume $n > 0$. Then $\frac{(1+fg)^n-1}{f}$ is defined to be $\Sigma_{i=1}^{n}\binom{n}{i}f^{i-1}g^i$. Then case (a) follows from D2 or D2', by induction on n. Thus

$$\langle f,g\rangle^n = \langle f,g\rangle^{n-1}\langle f,g\rangle = \langle f,\Sigma_{i=1}^{n-1}\binom{n-1}{i}f^{i-1}g^i\rangle\langle f,g\rangle$$

$$= \langle f,\Sigma_{i=1}^{n-1}\binom{n-1}{i}f^{i-1}g^i + g + \Sigma_{i=1}^{n-1}\binom{n-1}{i}f^ig^{i+1}\rangle$$

$$= \langle f,\Sigma_{i=1}^{n}[\binom{n-1}{i}+\binom{n-1}{i-1}]f^{i-1}g^i\rangle = \langle f,\Sigma_{i=1}^{n}\binom{n}{i}f^{i-1}g^i\rangle$$

$$= \langle f, \frac{(1+fg)^n - 1}{f} \rangle .$$

If n is negative and $(fg)^N = 0$ then $\frac{(1+fg)^n-1}{f}$ will be defined by the formal expansion $\Sigma_{i=1}^{N} \binom{n}{i} f^{i-1} g^i$ where $\binom{n}{i} = \frac{n(n-1)\ldots(n-i+1)}{i!} \in Z$. Working over $k[[X,Y]][X^{-1}]$ (X,Y indeterminates) the identity $\frac{(1+XY)^n-1}{X} + \frac{(1+XY)^{-n}-1}{X} + X(\frac{(1+XY)^n-1}{X})(\frac{(1+XY)^{-n}-1}{X}) = 0$ is easily checked. This is actually an identity in $k[[X,Y]]$. If we then truncate $(XY)^N = 0$ and replace X by f and Y by g we get the desired equation $(n>0)$

$$\langle f, \frac{(1+fg)^n-1}{f} \rangle \langle f, \frac{(1+fg)^{-n}-1}{f} \rangle = \langle f, 0 \rangle = 1 .$$

6. $\tilde{K}_2(k[X]/X^n)$ in finite characteristic. Let k be a ring of characteristic p (p a prime) (i.e. the sum $1 + 1 + \ldots + 1$ (p times) equals zero in k) and $n>2$. Note that every integer prime to p is a unit in k. Then the kernel of $\rho_1 : \Omega_k \to \tilde{K}_2(k[X]/X^n)$ is uncertain if $n = mp^r + 1$ and the kernel of $\rho_2 : k \to \tilde{K}_2 k[X]/X^n$ is uncertain if $n = mp^r$ $(m,p) = 1$.

Let $n = mp^r$ and consider

$$0 \to K \to \tilde{K}_2(k[X]/X^n) \to \tilde{K}_2(k[X]/X^{n-1}) \to 0$$

By theorem 1, K is generated by $\langle aX^{n-1},b\rangle$ and $\langle aX^{n-1},X\rangle$. We show that $\langle aX^{n-1},X\rangle = 0$ if $a = b^{p^r}$. Write $Y = X^m$ (so that $Y^{p^r} = 0$). Then $\langle aX^{n-1},X\rangle = \langle aX^{mp^r-1},X\rangle$ $= \langle am^{-1}X^{mp^r-m},X^m\rangle = \langle a'Y^{p^r-1},Y\rangle$. where $a' = am^{-1}$ is a p^rth power (m^{-1} lies in the prime field so is its own p^r-th root). Now $1 = \langle \alpha Y,1\rangle^{p^r} = \langle \alpha Y,\frac{(\alpha Y+1)^{p^r}-1}{\alpha Y}\rangle =$

$\langle \alpha Y,\frac{\alpha^{p^r}Y^{p^r}+1-1}{\alpha Y}\rangle = \langle \alpha Y,\alpha^{p^r-1}Y^{p^r-1}\rangle = \langle Y,\alpha^{p^r}Y^{p^r-1}\rangle =$ $\langle \alpha^{p^r}Y^{p^r-1},Y\rangle^{-1} = \langle(-\alpha)^{p^r}Y^{p^r-1},Y\rangle$. Thus $\langle b^{p^r}X^{n-1},X\rangle = 0$ for all $b \in k$, $n = mp^r$. This yields a homomorphism

$$\rho_2 : k/k^{p^r} \to \tilde{K}_2k[X]/X^{mp^r} .$$

The manuscript [2] suggests that (possibly with additional assumptions on k) the kernel of $\rho_2: k \to \tilde{K}_2k[X]/X^n$ is exactly k^{p^r}, but I have been unable to prove this. If $r=1$ we saw in the previous section that ρ_2 can be composed with a map $\tilde{K}_2k[X]/X^n \xrightarrow{\varphi} \Omega_k/n\Omega_k = \Omega_k$ such that $\varphi\rho_2(a) = da$. If k is a ring such that $a \in k^p \Leftrightarrow da = 0$ then the kernel of ρ_2 is exactly k^p. This is the case for k a field. It is also true for an integrally closed domain k with quotient field F. For if $da = 0$ in Ω_k, then $da = 0$ in Ω_F, $a = b^p$ $(b\in F)$ and integral closure implies that $b \in k$. (This argument was shown to me by Bill Heinzer). However it is not always true, even for k a domain. For let $k = F_p[X,Y]/(aX^{a-1}+Y^p)$, $F_p = Z/pZ$, X,Y indeterminates and $f = X^a + XY^p$. then $df = 0$. Let Y have weight $a-1$ and X have weight p . Then $\frac{\partial f}{\partial X} = aX^{a-1} + Y^p$ is homogeneous of weight $p(a-1)$. Thus k is graded.

The element f is homogeneous of weight pa . If $f = g^p$ then g must be homogeneous of weight a, but if $p > a > 2$ then k contains no elements of weight a . (Note that $f \neq 0$ in k . The only way $f = 0$ could occur in k would be for $X^a + XY^p$ to be a multiple in $F_p[X,Y]$ of $aX^{a-1} + Y^p$ and this is clearly not the case. Furthermore $aX^{a-1} + Y^p$ is irreducible if $1 < a < p$ so k is a domain). This example was shown to me by David Eisenbud.

If $r = 1$ and $k^p = \ker[d:k \to \Omega_k]$ then $d : k/k^p \to \Omega_k$ is a split inclusion because k/k^p and Ω_k are both vector spaces over F_p and d is F_p-linear. Therefore $\rho_2 : k/k^p \to \tilde{K}_2 k[X]/X^p$ is also a split inclusion. But this splitting does not seem particularly canonical.

<u>Theorem 3</u> If k is of characteristic p and every element of k has a pth root, then $\tilde{K}_2 k[X]/X^n = 0$.

Proof. Our assumption implies $\Omega_k = 0$ and $k/k^{p^r} = 0$. The case $n = 2$ is covered by the exact sequence $k/k^2 \to \tilde{K}_2 k[X]/X^2 \to \Omega_k \to 0$. Theorem 3 then follows by induction.

In the field case this theorem has been proved by Dennis-Stein [3] Lemma 3.4 and Morris [9]. Perhaps it is in [2] also. Rings which are not fields where Theorem 3 holds do exist, e.g. $k = Z/pZ[X, X^{\frac{1}{p}}, X^{\frac{1}{p^2}}, \ldots]$.

I do not know if there is a corresponding simplification in $\tilde{K}_2 k[X,Y]/(X,Y)^n$ when k is a perfect field (except if $n=2$ where Corollary 1.4 of [13] applies).

7. The case $k = Z$. If $k = Z$ the homomorphisms ρ_2 are all split inclusions. The following method of proof was suggested by Keith Dennis. This section is joint work by L. Roberts and S. Geller. We had a number of helpful conversations with C. Weibel.

First let $n = p^m$, a prime power, and consider the exact sequence

$$Z/p^m Z \xrightarrow{\rho_2} \tilde{K}_2 Z[X]/(X^{p^m}) \to \tilde{K}_2 Z[X]/(X^{p^m-1}) \to 0 .$$

We know that $\langle X^{n-1}, X\rangle$ is of order p^d, $d \leq m$. Let ζ be a primitive p^m-th root of unity. Then $pZ[\zeta] = (\pi^e)$ where $\zeta - 1 = \pi \in Z[\zeta]$ generates a prime ideal of $Z[\zeta]$ and $e = p^{m-1}(p-1)$. Define a homomorphism $Z[X]/(X^{p^m}) \to Z[\zeta]/\pi^{p^m}$ by sending X to π. This induces a map on the K_2's which sends $\langle X^{n-1}, X\rangle$ to $\langle \pi^{n-1}, \pi\rangle = \langle \pi^{n-1}, 1+\pi\rangle$ $= \{1+\pi^{n-1}, 1+\pi\}$. This is a generator of $K_2 Z[\zeta]/\pi^n$, by proposition 3.8 (f) of [3]. To apply this proposition we must check that $1 - \omega X - X^p = 0$ has no solution in $Z[\zeta]/\Pi = Z/pZ$, where $\omega\Pi^e = p$. The sum of coefficients in $(X-1)^e$ is zero. Let Φ be the minimal polynomial of ζ. Φ is of degree e and has p terms each with coefficient 1. Mod p we have $X^{p^m}-1 = (X-1)^{p^m}$ so $\Phi(X) \equiv (X-1)^e \bmod p$. Thus $\psi(X) = (X-1)^e - \Phi(X)$ is a polynomial that vanishes in Z/pZ, so all its coefficients are divisible by p. The sum of the coefficients of $\psi(X)$ is $-p$ and $\zeta \equiv 1 \bmod \Pi$, so we have $\omega^{-1} = p^{-1}\Pi^e = p^{-1}\psi(\zeta) \equiv -1 \bmod \Pi$. But $1 + X - X^p$ clearly has no solutions in $Z[\zeta]/\Pi = Z/pZ$ so we can use proposition 3.8.

By Theorem 5.1 of [3], $K_2Z[\zeta]/(\Pi^n)$ is cyclic of order p. Thus there is a homomorphism from $\tilde{K}_2Z[X]/X^{p^m}$ to Z/pZ which sends $\langle X^{n-1},X\rangle$ to the generator. By the fundamental theorem of abelian groups we get that $\langle X^{n-1},X\rangle$ must generate a direct summand of $\tilde{K}_2Z[X]/X^{p^m}$, so we get a split exact sequence

$$0 \to Z/p^dZ \to \tilde{K}_2Z[X]/X^{p^m} \to \tilde{K}_2Z[X]/X^{p^m-1} \to 0$$

where $1 \leq d \leq m$.

We now use the transfer map to handle composite n. Let $n = p_1^{a_1} p_2^{a_2} \cdots p_r^{a_r}$, p_i distinct primes, $a_i \geq 1$. Then set $p = p_i$, $a = a_i$ and $n = p^a m$ with $(m,p) = 1$. We have an inclusion $A = Z[t]/t^{p^a} \xrightarrow{i} Z[X]/X^n = B$, given by

$t \to X^m$. B is a free A-module, with basis $1, X, \ldots, X^{m-1}$, so the transfer map $i_* : \tilde{K}_2(B) \to \tilde{K}_2(A)$ exists, and $i_* i^* = m$. But $\langle X^{mp^a - m}, X^m\rangle = \langle X^{mp^a-1}, X\rangle^m$. Applying i_* to both sides we get $\langle t^{p^a-1}, t\rangle^m = [i_* \langle X^{n-1}, X\rangle]^m$. Under the composition $\tilde{K}_2(B) \to \tilde{K}_2(A) \to Z/pZ$ (where the second map is as obtained in the prime power case) we get $\langle X^{n-1}, X\rangle$ mapping to a generator (since m is a unit in Z/pZ). Thus $\langle X^{n-1}, X\rangle$ has order at least $p_1 p_2 \ldots p_r$ and must generate a direct summand of $\tilde{K}_2(B)$. The map $\tilde{K}_2 Z[X]/X^n \to \tilde{K}_2 Z[X]/X^{n-1}$ therefore splits.

Now we show that if $n = p^m$ the order of $\langle X^{n-1}, X\rangle$ in $\tilde{K}_2 Z[X]/X^n$ is exactly p^m . By Theorem 5.1 of [3] we can choose $\ell \geq p^m$ so that $\tilde{K}_2 Z[\zeta]/\Pi^\ell$ is cyclic of order p^m . Then we have a commutative diagram (with vertical arrows onto)

$$\begin{array}{ccc} \tilde{K}_2 Z[X]/X^\ell & \to & K_2 Z[\zeta]/\pi^\ell = Z/p^m Z \\ \downarrow & & \downarrow \\ \tilde{K}_2 Z[X]/X^{p^m} & \to & K_2 Z[\zeta]/\Pi^{p^m} = Z/pZ \end{array}$$

If $\langle X^{n-1}, X\rangle$ is of order p^d in $\tilde{K}_2 Z[X]/X^{p^m}$ then (by the splitting proved in the preceeding paragraph) we can lift it to an element $a \in \tilde{K}_2 Z[X]/X^\ell$ which is of order p^d . But a maps to a generator in Z/pZ, so it maps to a generator in Z/p^mZ. Thus $p^d \geq p^m$. But $d \leq m$ so $d = m$ as required, and ρ_2 is an inclusion if n is a prime power. We get ρ_2 an inclusion in the composite n case also because the transfer map argument now shows that if $n = p_1^{a_1} \ldots p_r^{a_r}$ then $\langle X^{n-1}, X\rangle$ has order at least $p_1^{a_1} \ldots p_r^{a_r} = n$. Thus we have proved:

Theorem 4. $\tilde{K}_2 Z[X]/X^n = \bigoplus_{i=2}^{n} Z/iZ$.

If S is a multiplicative set in Z then the same argument shows that $\tilde{K}_2 Z_S[X]/X^n = (\bigoplus_{i=2}^{n} Z/iZ) \otimes_Z Z_S$, i.e. only the primes in S are killed off. In particular we have computed $\tilde{K}_2 R[X]/X^n$ for R any ring between Z and Q, since all such rings are localizations.

8. Some explicit formulas. We saw in §7 that the homomorphisms ρ_2 are split inclusions if $k = Z$. In this section I will show how explicit formulas for the splitting can be obtained, at least if n is small. Let $R = Z[X]/X^2$ and $I = XR$. The generators of $D(R,I)$ are $\langle aX+b, cX\rangle$ and $\langle cX, aX+b\rangle$, $a,b,c \in Z$. In $D(R,I)$ we have $1 = \langle cX, 1\rangle^{-b} = \langle cX, -b + \frac{b(b+1)}{2} cX\rangle$ (by lemma 2). Hence $\langle cX, aX+b\rangle = \langle cX, aX+b\rangle\langle cX, 1\rangle^{-b} = \langle cX, aX+b\rangle\langle cX, -b + \frac{b(b+1)}{2} cX\rangle =$
$\langle cX, aX + \frac{b(b+1)}{2} cX - b^2 cX\rangle = \langle cX, (a - \frac{b(b-1)}{2} c)X\rangle$
$= \langle c(a - \frac{b(b-1)}{2} c)X, X\rangle$. A similar computation shows that

$<aX+b,cX> = <aX+b,cX><-1,cX>^b = <(a - \frac{b(b+1)}{2}c)X,cX>$ $= <c(a - \frac{b(b+1)}{2}c)X,X>$. Hence our splitting φ for ρ_2 is given by $\varphi<aX+b,cX> = c(a + \frac{b(b+1)}{2}c)$ and $\varphi<cX,aX+b> = c(a + \frac{b(b-1)}{2}c)$. The c inside the parentheses can be omitted since $c^2 \equiv c \bmod 2$.

Now consider the case $n = 3$. We have an exact sequence

$$0 \to Z/3Z \overset{\rho_2}{\to} \tilde{K}_2Z[X]/X^3 \overset{\pi}{\to} \tilde{K}_2Z[X]/X^2 \to 0 .$$

By D1, $<X,-X> \in \tilde{K}_2Z[X]/X^3$ is of order 2. (In fact $<X,-X>$ is of order 2 in $\tilde{K}_2Z[X]/X^n$, for any n). And $\pi<X,-X>$ = non-trivial element of $\tilde{K}_2\ Z[X]/(X^2) = Z/2Z$. Thus

$$\tilde{K}_2(Z[X]/X^3) = (\text{image } \rho_2) \oplus Z/2Z$$

($<X,-X>$ being a generator of $Z/2Z$). Now we take a generator of the form $<aX+bX^2,c+dX+eX^2>$ and find its projection onto image ρ_2 under this direct sum decomposition. Note that $<aX^2,b> = 0$ (since this is the image under ρ_1 of $adb \in \Omega_Z = 0$), and $<aX,-bX> = <-aX^2,+b><abX,-X>$ $= <abX,-X>$ (by D3) and $<aX,-X> = <X,-X>^a$. Then $<aX+bX^2,c+dX+eX^2> = <aX,c+dX+eX^2><bX^2,c+dX+eX^2>$ $= <aX,c+dX><aX,eX^2><bX^2,c><bX^2,dX><bX^2,eX^2>$ $= <aX,c+dX><(bd-ae)X^2,X>$. Now we tackle the first term, working $\bmod <X,-X>$. Thus $<aX,c+dX> = <aX,c+dX><aX,-dX>$

$= \langle aX, c-acdX^2\rangle = \langle aX,c\rangle\langle aX,-acdX^2\rangle = \langle aX,c\rangle\langle a^2cdX^2,X\rangle$.

Now we consider $\langle aX,c\rangle$. By lemma 2 we have

$$\langle aX,c\rangle\langle aX,1\rangle^{-c} = \langle aX,c\rangle\langle aX,-c + \frac{c(c+1)}{2}aX - \frac{c(c+1)(c+2)}{3!}a^2X^2\rangle$$

$$= \langle aX,(\frac{c(c+1)}{2}a-ac^2)X + (\frac{a^2c^2(c+1)}{2} - \frac{c(c+1)(c+2)}{3!}a^2)X^2\rangle$$

$$= \langle aX,(\frac{c(c+1)}{2}a-ac^2)X\rangle\langle aX,(\frac{a^2c^2(c+1)}{2} - \frac{a^2c(c+1)(c+2)}{3!})X^2\rangle$$

$$= \langle(\frac{a^3c(c+1)(c+2)}{3!} - \frac{a^3c^2(c+1)}{2})X^2,X\rangle$$ (again factoring out by $\langle X,-X\rangle$) .

$$= \langle a(\frac{c(c+1)(c+2)}{3!} - \frac{c^2(c+1)}{2})X^2,X\rangle$$

(since $a^3 \equiv a \bmod 3$). Putting everything together we get $\langle aX+bX^2,c+dX+eX^2\rangle = \langle(bd-ae+a^2cd + \frac{ac(c+1)(c+2)}{3!} - \frac{ac^2(c+1)}{2})X^2,X\rangle$ (mod $\langle X,-X\rangle$). Thus the splitting φ for ρ_2 is given by $\varphi\langle aX+bX^2,c+dX+eX^2\rangle$ $= bd-ae+a^2cd + \frac{a}{3}(c-c^3)$ and $\varphi\langle c+dX+eX^2,aX+bX^2\rangle$ $= -\varphi\langle -aX-bX^2,-c-dX-eX^2\rangle = -bd+ae-a^2cd + \frac{a}{3}(c^3-c)$.

Now let $n = 4$. We have an exact sequence

$$0 \to Z/4Z \xrightarrow{\rho_2} \tilde{K}_2Z[X]/(X^4) \xrightarrow{\pi} \tilde{K}_2Z[X]/X^3 \to 0$$

$\tilde{K}_2Z[X]/(X^3) = Z/3Z \oplus Z/2Z$ with generators $\langle X^2,X\rangle$ and $\langle X,-X\rangle$. In $\tilde{K}_2Z[X]/X^4$ it follows from D2 and D3 that $\langle aX^2,bX\rangle$ is linear in a and b and that $\langle aX^2,bX\rangle = \langle abX^2,X\rangle$ The proof that $\langle X^2,X\rangle$ is of order 3 in $\tilde{K}_2Z[X]/(X^3)$ is also valid in $\tilde{K}_2Z[X]/(X^4)$. We now find the projection φ of $K_2Z[X]/X^4$ onto image $\rho_2 = Z/4Z$ by setting $\langle aX^2,X\rangle$ and $\langle X,-X\rangle$ equal to one where they occur.

First consider generators of $\tilde{K}_2 Z[X]/X^4$ of the following type: $\langle aX+bX^2+cX^3, eX+fX^2+gX^3\rangle$. We have $\langle aX+bX^2+cX^3, eX+fX^2+gX^3\rangle = \langle aX, gX^3\rangle\langle cX^3, eX\rangle\langle aX+bX^2, eX+fX^2\rangle =$ $\langle (ce-ag)X^3, X\rangle\langle bX^2, fX^2\rangle\langle aX, fX^2\rangle\langle bX^2, eX\rangle\langle aX, eX\rangle$ $= \langle (ce+2bf-ag)X^3, X\rangle\langle aX, eX\rangle$. Now we concentrate on $\langle aX, eX\rangle$. By lemma 2, $\langle X, -X\rangle^a = \langle aX - \frac{a(a-1)}{2}X^3, -X\rangle$ $= \langle aX, -X\rangle\langle \frac{a(a-1)}{2}X^3, X\rangle$, and $\langle aX, -X\rangle^{-e} =$ $\langle aX, eX + \frac{e(e+1)}{2}aX^3\rangle = \langle aX, eX\rangle\langle aX, \frac{e(e+1)}{2}aX^3\rangle =$ $\langle aX, eX\rangle\langle -\frac{a^2e(e+1)}{2}X^3, X\rangle$. Altogether we get $\langle aX, eX\rangle = \langle \frac{ae(2a+ae-1)}{2}X^3, X\rangle$ (in the projection onto image ρ_2 given by the above splitting for π), and the image of $\langle aX+bX^2+cX^3, eX+fX^2+gX^3\rangle$ under our proposed splitting φ for ρ_2 is $ce + 2bf - ag + \frac{ae(2a+ae-1)}{2}$.

The general definition of φ is then $\varphi\langle aX+bX^2+cX^3, d+eX+fX^2+gX^3\rangle = \varphi[\langle aX+bX^2+cX^3, d+eX+fX^2+gX^3\rangle$ $\langle aX+bX^2+cX^3, 1\rangle^{-d}]$ and $\varphi\langle d+eX+fX^2+gX^3, aX+bX^2+cX^3\rangle =$ $-\varphi\langle -aX-bX^2-cX^3, -d-eX-fX^2-gX^3\rangle$. First find one symbol for $\langle aX+bX^2+cX^3, 1\rangle^{-d}$ by lemma 2. Then use D2 to obtain one symbol for the expression in square brackets (eliminating the constant term) and apply the formula for φ with no constant term. The resulting explicit formula is written down in [10a], but is so messy that I will not give it here. Note however that it contains expressions of the form $d(d+1)/2$, $d(d+1)(d+2)/3!$ and $d(d+1)(d+2)(d+3)/4!$ all of which are integers if $d \in Z$.

Because of these denominators our formulas do not give a splitting for ρ_2 except for $k = Z$ (or localizations thereof). In fact, let $k = Z[i]$ the Gaussian integers. An unpublished computation of Dennis and Weibel shows that $\tilde{K}_2 k[X]/X^2 = Z/2Z \oplus Z/2Z \oplus Z/4Z$, whereas $k/2k = Z/2Z \oplus Z/2Z$ and $\Omega_k = Z/2Z \oplus Z/2Z$. Thus in this case ρ_2 is an inclusion but does not split.

REFERENCES

1. H. Bass, Algebraic K-Theory, Benjamin, New York, 1968.

2. S. Bloch, Relative K_2 for truncated polynomial rings. Manuscript

(a) S. Bloch, K_2 of artinian Q-algebras, Comm. in Alg. 5 (1975), 405-428.

(b) K. Dennis, Differentials in Algebraic K-theory (unpublished).

3. R.K. Dennis and M. Stein, K_2 of discreet valuation rings, Advances in Math. 18 (1975), 182-238.

4. R.K. Dennis and M. Stein, The functor K_2, a survey of computations and problems, Lecture Notes in Math. vol. 342, Springer-Verlag, Berlin and New York, 1973, pp 243-280.

5. S. Geller and L.G. Roberts, Further results on excision for K_1 of algebraic curves. Preprint, 1977.

(b) S. Gersten, Some exact sequences in the higher K-theory of rings, Lecture Notes in Math. vol. 341, Springer-Verlag, Berlin and New York, 1973, pp 211-243.

6. W. van der Kallen, Le K_2 des nombres duaux. C.R. Acad. Sc. Paris (1971), 1204-1207.

7. H. Matsumura, Commutative Algebra, Benjamin, New York, 1970.

8. H. Maazen and J. Stienstra, A presentation for K_2 of split radical pairs, J. of Pure and Applied Alg. 10 (1977), 271-294.

9. R. Morris, Derivations of Witt vectors with application to K_2 of truncated polynomial rings and Laurent series. Preprint, 1977.

10. L. G. Roberts, SK_1 of n lines in the plane, Trans. Amer. Math. Soc. 222 (1976), 353-365.

10(a) L. G. Roberts, K_2 of some truncated polynomial rings, Queen's Mathematical Preprint No. 1978-17 . Queen's University, Kingston, Ontario.

11. R. Swan, Excision in Algebraic K-Theory, J. of Pure and Applied Alg., 1 (1971), 221-252.

12. T. Vorst, Polynomial extensions and excision for K_1, preprint 63 (1977) Utrecht University.

13. C. A. Weibel, K_2, K_3 and nilpotent ideals, preprint, 1978.

IS THE BRAUER GROUP GENERATED BY CYCLIC ALGEBRAS?

by

Robert L. Snider

Virginia Polytechnic Institute and State University
Blacksburg, Va. 24061

and

The Institute for Advanced Studies
The Hebrew University
Jerusalem, Israel

Is the Brauer Group Generated by Cyclic Algebras?

by

Robert L. Snider

Let k be a field containing a primitive nth root of unity with n prime to the characteristic of k. The power norm residue map $R_{n,k}$ is a homomorphism from $K_2(k)$ to $Br_n(k)$ where $Br_n(k)$ is the subgroup of the Brauer group which is annihilated by n. If {a,b} is a symbol, then $R_{n,k}\{a,b\}$ is the n^2 dimensional algebra generated by x and y with $x^n = a$, $y^n = b$, and $xyx^{-1} = \omega y$ where ω is a primitive nth root of unity [15]. $R_{n,k}$ is clearly split by the cyclic field $k(\sqrt[n]{y})$ and hence is similar to a cyclic algebra. The image then of $R_{n,k}$ is the algebras which are similar to a product of cyclic algebras. It is unknown if $R_{n,k}$ is surjective. Perhaps the best positive result is a theorem of Rosset that division algebras of degree p are in the image [18].

Since every division algebra is similar to a crossed product, to show $R_{n,k}$ is surjective, it suffices to show crossed products are in the image of $R_{n,k}$. We show that to prove crossed products with group G are in the image, it is sufficient to prove that the fixed field of a certain rational function field with G acting on it is purely transcandental.

Our main technical device is the construction of a generic crossed product with group G. This is the quotient division algebra of a certain group ring. It is shown that it suffices to show the generic crossed produc is similar to a product of cyclics in order to show that every crossed produ with group G is similar to a product of cyclic algebras. This is done in section 1.

In section 2, we apply a theorem of Bloch [4] to show that if the enter of the generic crossed product is purely transcendental, then it s similar to a product of cyclics. The center is the fixed field of a urely transcendental extension of k with G acting on it. This implies hat if the title question is false for a field containing an algebraically losed field, then the center of a generic crossed product s a unirational field which is not rational. Very few such examples are nown [6, 16].

In sections 3 and 4, we make detailed calculations of the center in ertain special cases. We show that if D is 16 dimensional over its enter and the center contains a primitive 4th root of 1, then D is imilar to a product of cyclics. Similar results are obtained for rossed products with group D_n, n odd.

In our last section, we list several open questions.

We were inspired by an interesting but false paper of Rosset [17].

We are using the word"similar"to mean two algebras represent the ame element in the Brauer group. We will always assume that fields re infinite. We will denote the ring of quotients of a ring R by QR.

1. Generic Crossed Products

Let G be a finite group. We form a free presetation

$$1 \to R \to F \to G \to 1.$$

Factoring out R', we obtain the free abelian extension [11,§9.5]

$$1 \to R/R' \to F/F' \to G \to 1$$

We shall always write $\overline{R} = R/R'$ and $\overline{F} = F/R'$. $\overline{R}$ is a G-module called the relation module of G. It depends on the presentation but different presentations have closely related relation modules. See [10] for details.

If $1 \to A \to E \to G \to 1$ is an extension with A abelian, then there exists $\phi: \overline{F} \to E$ such that the diagram commutes

$$\begin{array}{ccccccccc} 1 & \to & \overline{R} & \to & \overline{F} & \to & G & \to & 1 \\ & & & & \phi\downarrow & & \| & & \\ 1 & \to & A & \to & E & \to & G & \to & 1. \end{array}$$

This follows immediately from the freeness of F. Clearly $\phi(\overline{R}) \subseteq A$.

Let k be a field and $B = (K,G,f)$ a crossed product with group G such that $k \subseteq Z(B)$. (See [13] for definitions and basic results on crossed products). We let X_g denote elements of B such that $X_g^{-1} tX_g = t^g$ for all $t \in K$. The group E generated by K^* and the X_g gives rise to an extension

$$1 \to K^* \to E \to G \to 1$$

by sending X_g to g. Therefore we have a homomorphism $\phi: \overline{F} \to E$ which extends linearly to a homorphism from the group ring $k[\overline{F}]$ to B.

$\overline{F}$ is a torsion free abelian-by-finite group and hence $k[\overline{F}]$ has a division ring of fractions $Qk[\overline{F}]$ (See [8] or [14]). We shall call k[$\overline{F}$] the <u>generic crossed product</u> with group G. If F has at least o generators (and we shall always assume this), then $\overline{R}$ is a faithful module [10, p.8]. This implies that Qk[$\overline{F}$] is a crossed product th maximal subfield Qk[$\overline{R}$] and group G. To see this we note that the nter of Qk[$\overline{F}$] is the ring of invariants (fixed points) Qk[$\overline{R}$]^G and nce Qk[$\overline{R}$] has dimension $|G|$ over Qk[$\overline{R}$]^G. Also Qk[$\overline{F}$] has dimension | over Qk[$\overline{R}$]. Therefore Qk[$\overline{R}$] is a maximal subfield.

We mention that there is already a notion of a generic abelian crossed oduct [3]. This is a different object from our construction even in e abelian case.

Our idea is that to prove results about crossed products, it suffices prove them about the generic crossed product and then transfer the sults by specialization. In order to do this, we must have sufficienty ny speciazations. This is the content of our first result.

We let B = (K,G,f) a crossed product with group G and ϕ a specialization nstructed as above.

<u>Theorem 1</u>. If $a \neq 0 \in k[\overline{R}]$, <u>then there is a homorphism</u> <u>: k[$\overline{F}$] $\to$ B such that $\psi(a) \neq 0$</u>.

We first consider the case that F is generated by y_g, $g \neq 1$, $g \in G$ d the presentation is obtained by sending y_g to g. In this case 1 and e y_g is a Schreier system and hence $Z_{g,h} = y_g y_h y_{gh}^{-1}$, $g \neq 1$, $h \neq 1$ is a t of free generators for $\overline{R}$ where we set $y_{gh} = 1$ if $gh = 1$. (See [12] r details).

Let $\phi(y_g) = b_g$ and $\phi(Z_{g,h}) = C_{g,h}$. We obtain a new specialization defining $\psi(y_g) = t_g b_g$ where $t_g \in K^*$ and $t_1 = 1$.

Then $\phi(Z_{g,h}) = t_g t_n^{g^{-1}} t_{gh}^{-1} C_{g,h}$.

We may clearly assume that a is a polynomial in the $Z_{g,h}$. We induct on the number of $Z_{g,h}$ in a. If this number is one, the result follows since a would then have an infinite number of solutions.

Suppose then that $Z_{r,s}$ appears in a. We write $a(Z_{g,h}) = \sum_{i=0}^{n} a_i(Z_{g,h}) Z_{r,s}^i$ where $Z_{r,s}$ does not appear in a_i.

Suppose to the contrary that a vanishes for all specializations. We let $t_g = 1$ for $g \neq s$ and substitute the $C_{g,h}$ to obtain

$$a(t_g t_h^{g^{-1}} t_{gh}^{-1} C_{g,h}) = \sum_{i=0}^{n} a_i(t_g t_h^{g^{-1}} t_{gh}^{-1} C_{g,h})(t_r t_s^{r^{-1}} t_{rs}^{-1})^i C_{r,s}^i$$

This is a finite Laurent series in t_s^g which vanishes for all $t_s \in K^*$ and hence is identically 0 [5, p.144]. The coefficient of $(t_s^{r^{-1}})^i$ is $a_i(t_g h_h^{g^{-1}} t_{gk}^{-1} C_{g,h}) t_r^i t_{rs}^{-i} C_{r,s}^i$ This must also be identically 0 and hence letting $t_s = 1$, we obtain $a_i(C_{g,h}) = 0$. But $C_{g,h}$ could be any specializatio and hence $a_i = 0$ by induction and hence $a = 0$, a contradiction.

Now let F be free on $\{x_i\}$ and F_1 be free on $\{x_i\} \cup \{y_i\}$. If $1 \to R \to F \overset{\pi}{\to} G \to 1$ is a presentation and $\overline{\pi} = F_1 \to G$ extends π, then we may change the generators of F_1 so that F_1 is free on $\{x_i\} \cup \{z_j\}$ with $\overline{\pi}(z_j) = 1$. We have $\overline{R}_1 = \overline{R} \oplus \langle z_j^g : g \in G\rangle$. $\langle z_j^g\rangle$ is a free G-module with basis z_j. Suppose the theorem holds for $k[\overline{F}]$. If $a \in k[\overline{R}_1]$, we write a as a polynomial in the z_j^g with coefficients in $k[\overline{R}]$. By the theorem, we can find a specialization such that the coefficients do not vanish. Now by applying proposition 2, p.144 of [5], e may specialize the Z_j to obtain a nonzero specialization of a. Conversely f the theorem is true for $k[\overline{F}_1]$, it is clearly true for $k[\overline{F}]$.

Now if $1 \to R \to F \to G \to 1$ is any presentation, we may enlarge F to $_1$ and then cut down to F_2 where F_2 is the special presentation for which e have proved the theorem. Therefore the theorem is proved for all resentations.

Corollary 1. If $a_1,\ldots,a_n$ are nonzero elements of $k[\overline{R}]$, then there s a homomorphism $\psi: k[\overline{F}] \to B$ such that $\psi(a_i) \neq 0$ for all i.

Proof: Apply the theorem to $\pi\, a_i$.

Theorem 2. If the generic crossed product $Qk[\overline{F}]$ with group G is imilar to a product of cyclic algebras, then every crossed product ith group G and center containing k is similar to a product of cyclic lgebras.

This theorem is proved by using the ideas Amitsur developed to ransfer results from the generic division algebra to arbitrary division lgebras [2].

We shall need only a special case of this theorem. We shall therefore ssume that k contains a primitive nth root of unity, all cyclic algebras nvolve groups whose orders divide n, and that n is prime to the haracteristic of k.

Proof of the special case. Since $Qk[\overline{F}]$ is a division algebra, there s an integer m such that $M_m(Qk[\overline{F}]) = A_1 \otimes_Z \cdots \otimes_Z A_r$ where Z is the enter of $Qk[\overline{F}]$ and each A_i is cyclic. Since $Z = Qk[\overline{R}]^G$ is the field f fractions of $S = k[\overline{R}]^G$, the center of $k[\overline{F}]$, it follows that $Qk[\overline{F}]$ is ormed by inverting the nonzero elements of S. Under our hypothesis, for each A_i, there is an integer n_i and elements a_i and b_i in A_i such that a^{n_i} and $b^{n_i} \in Z$, $a_ib_i = \omega_ib_ia_i$ where ω_i is a primitive n_ith root of 1, and A is spanned by the products a^kb^l. By multiplying by an element of S, we may assume that a_i and $b_i \in M_m(k[\overline{F}])$. Let C_i be the n_i^2 dimensional S algebra generated by a_i and b_i.

If B is a crossed product with group G such that $k \subseteq Z(B)$, then by the corollary, there is a homomorphism ψ: $k[\overline{F}] \to B$ such that $\psi(a_i^{n_i}) \neq 0$ and $\psi(b_i^{n_i}) \neq 0$ for all i. ψ extends to a homomorphism from $M_m(k[\overline{F}])$ to $M_m(B)$. Let $K = Z(B)$. $K\psi(C_i)$ is a K algebra generated by $\psi(a_i)$ and $\psi(b_i)$. Furthermore $\psi(a_i)\psi(b_i) = \omega_i\psi(b_i)\psi(a_i)$ and $\psi(a_i)^{n_i} \neq 0$, $\psi(b_i)^{n_i} \neq 0 \in K$. Therefore $K\psi(C_i)$ is similar to a cyclic algebra and has dimension n_i^2 over K. Also the center of $K\psi(C_i)$ is K. Now $K\psi(C_1)$ $K\psi(C_2)$ ... $K\psi(C_r)$ must be a tensor product since they centralize each other and are central simple A dimension count now shows that $M_m(B) = K\psi(C_1) \otimes_K \cdots \otimes_K \psi(C_r)$.

§2. Bloch's Theorem

Bloch [4] has shown that the Kernels and cokernels of the power norm residue maps

$$R_{n,k}: K_2(k)/nK_2(k) \to Br_n(k)$$

and $R_{n,k(X_1,\ldots,X_m)} : K_2(k(x_1,\ldots,X_m))/nK_2k(x_1,\ldots,X_m) \to Br_n(k(X_1,\ldots,X_m))$ are isomorphic.

These results all require that k contain a primitive nth root of ity and that n is prime to the characteristic of k.

If k is algebraically closed, a number field, or a function field er a finite field, then the top map is surjective. Therefore we can mbine theorem 1.2 and Bloch's theorem to obtain

Theorem 1. *If* $Qk[\overline{R}]^G$ *is purely transcendental over k, k contains primitive nth root of unity where n = |G|, n is prime to the aracteristic of k, and k is either algebraically closed, a number eld, or a function field over a finite field, then a crossed product th group G whose center contains k is similar to a product of cyclic gebras*.

We remark that if k is algebraically closed, then $k[\overline{R}]^G$ is unirational- at is, it is contained in a purely transcendental extension. There are ry few such fields known [6, 16]. All proofs of the existence of such elds require considerable algebraic geometry. It would be also very teresting if these fields turned out not to be purely transcendental.

3. *The Klein 4-group*

In this section we shall prove

Theorem 1. *If* $G = Z_2 \times Z_2$, *then* $Qk[\overline{R}]^G$ *is purely transcendental er k*.

This has as as immediate consequence the

Theorem 2. If D is a division algebra 16 dimensional over its center K and $\sqrt[4]{1} \in K$, then D is similar to a product of cyclic algebras.

Proof: In this case D is a crossed product with group $Z_2 \times Z_2$[1].

We remark that if char D = 2, then D is actually similar to a cyclic algebra [1].

Proof of theorem 1. We first prove the theorem in the special case that F is free on x and y and the presentation sends x to (1,0) and y to (0,1). 1,x,y,xy is a Schreier system and hence $\overline{R}$ is free on

$$t_1 = x^2,\ t_2 = y^2,\ t_3 = yxy^{-1}x^{-1},\ t_4 = xyxy^{-1},\ t_5 = xy^2x^{-1}.$$

Let α be conjugation by x and β conjugation by y. Routine calculation give the following table:

$t_1 = x^2$	$\alpha(t_1) = t_1$	$\beta(t_1) = t_3t_4$
$t_2 = y^2$	$\alpha(t_2) = t_5$	$\beta(t_2) = t_2$
$t_3 = yxy^{-1}x^{-1}$	$\alpha(t_3) = t_1^{-1}t_4$	$\beta(t_3) = t_2t_3^{-1}t_5^{-1}$
$t_4 = xyxy^{-1}$	$\alpha(t_4) = t_1t_3$	$\beta(t_4) = t_1t_2^{-1}t_3t_5$
$t_5 = xy^2x^{-1}$	$\alpha(t_5) = t_2$	$\beta(t_5) = t_5$

Claim: The field of invariants of α is

$$L = k(t_1, t_3t_4, t_4 + t_1t_3, t_2 + t_5, t_2t_4 + t_1t_3t_5).$$

s clearly invariant. Also $[L(t_4):L] = 2$ since trace $(t_4) \in L$ and m $(t_4) \in L$. But $t_1, t_4 \in L(t_4)$ implies $t_3 \in L(t_4)$. One can solve t_2 and t_5 since $\begin{vmatrix} 1 & 1 \\ t_4 & t_1t_3 \end{vmatrix} \neq 0.$

refore $L(t_4) = Qk[\overline{R}]$ and hence L is a fixed field of codimension 2 the claim is established.

cts on L. $Qk[\overline{R}]^G$ is the invariants of L under β.

name the variables of L, $a_1 = t_1$, $a_2 = t_3t_4$, $a_3 = t_2 + t_5$, $= t_2t_4 + t_1t_3t_5$, $a_5 = t_4 + t_1t_3$.

We compute the action of β.

$$\beta(a_1) = \beta(t_1) = t_3t_4 = a_2$$

$$\beta(a_2) = \beta(t_3t_4) = t_2t_3^{-1}t_5^{-1}t_1t_2^{-1}t_3t_5 = t_1 = a_1$$

$$\beta(a_3) = \beta(t_2 + t_5) = t_2 + t_5 = a_3$$

$$\begin{aligned}\beta(a_4) &= \beta(t_2t_4 + t_1t_3t_5) = t_2t_1t_2^{-1}t_3t_5 + t_3t_4t_2t_3^{-1}t_5^{-1}t_5 \\ &= t_1t_3t_5 + t_2t_4 = a_4\end{aligned}$$

$$\beta(a_5) = \beta(t_4)+\beta(t_1t_3) = t_1t_2^{-1}t_3t_5 + t_3t_4t_2t_3^{-1}t_5^{-1} = t_1t_2^{-1}t_3t_5 + t_2t_4t_5^{-1}$$

must compute $\beta(a_5)$. We have

$$\begin{aligned}(t_2^{-1} + t_5^{-1})a_4 &= (t_2^{-1} + t_5^{-1})(t_2t_4 + t_1t_3t_5) = t_4 + t_1t_3 + t_1t_2^{-1}t_3t_5 + t_2t_4t_5^{-} \\ &= a_5 + \beta(a_5).\end{aligned}$$

refore $\beta(a_5) = (t_2^{-1} + t_5^{-1})a_4 - a_5$

Now $t_2 + t_5 = a_3$

$$t_2t_4 + t_1t_3t_5 = a_4$$

By Cramer's rule, $t_2 = \dfrac{a_3t_1t_3 - a_4}{t_1t_3 - t_4}$ and $t_5 = \dfrac{a_4 - a_3t_4}{t_1t_3 - t_4}$

Hence $t_2^{-1} + t_5^{-1} = \dfrac{t_1t_3 - t_4}{a_3t_1t_3 - a_4} + \dfrac{t_1t_3 - t_4}{a_4 - a_3t_4} = \dfrac{(t_1t_3 - t_4)(t_1t_3 - t_4)a_3}{-a_4^2 - a_3^2t_1t_3t_4 + (t_1t_3 + t_4)a_3a_4}$

$$= \frac{(t_1t_3 - t_4)^2a_3}{-a_4^2 - a_3^2a_1a_2 + a_5a_3a_4} = \frac{[(t_1t_3 + t_4)^2 - 4t_1t_3t_4]a_3}{-a_4^2 - a_3^2a_1a_2 + a_3a_4a_5}$$

$$= \frac{(a_5^2 - 4a_1a_2)a_3}{-a_4^2 - a_3^2a_1a_2 + a_3a_4a_5}$$

Therefore

$$\beta(a_5) = \frac{-4a_1a_2a_3a_4 + a_5(a_4^2 + a_3^2a_1a_2)}{-a_4^2 - a_3^2a_1a_2 + a_3a_4a_5}$$

Denote a_5 by Z. β gives an automorphism of $k(a_1,a_2,a_3,a_4)$

Now

$\beta(Z) = \dfrac{aZ + b}{cZ + d}$ where a,b,c,d are the coefficients in the above formula.

For $e \in k(a_1,a_2,a_3,a_4)$, we have

$$\beta(Z + e) = \frac{aZ + b}{cZ + d} + \beta(e) = \frac{(a + \beta(e)c)Z + b + \beta(e)d}{cZ + d}$$

Let $\beta(e) = -\frac{a}{c}$, then

$$\beta(Z + e) = \frac{b - \frac{a}{c}d}{c(Z + e) + d - ce}$$

If we replace Z by $\overline{Z} = Z + e$, we have

$$\beta(\overline{Z}) = \frac{b - \frac{a}{c}d}{c\,\overline{Z} + d - ce}$$

Since $e = \frac{-\beta(a)}{\beta(c)}$, $d - ce = -a_4^2 - a_3^2a_1a_2 + a_3a_4\left(\frac{a_4^2 + a_3^2a_1a_2}{a_3a_4}\right) = 0.$

$$\text{Hence } \beta(\overline{z}) = \frac{b - \frac{a}{c}d}{c\overline{z}} = \frac{-4a_1a_2a_3a_4 - \frac{(a_4^2 + a_3^2a_1a_2)(-a_4^2 - a_3^2a_1a_2)}{a_3a_4}}{a_3a_4\overline{z}}$$

$$= \frac{-4a_1a_2a_3^2a_4^2 + (a_4^2 + a_3^2a_1a_2)^2}{a_3^2a_4^2\overline{z}} = \frac{(a_4^2 - a_3^2a_1a_2)^2}{(a_3a_4)^2\overline{z}}$$

Replace $\overline{z}$ by $\frac{a_3a_4}{a_4^2 - a_3^2a_1a_2}\,\overline{z} = \overline{\overline{z}}$

$$\text{Then } \beta(\overline{\overline{z}}) = \beta\left(\frac{a_3a_4}{a_4^2 - a_3^2a_1a_2}\right)\beta(\overline{z}) = \frac{a_3a_4}{a_4^2 - a_3^2a_1a_2}\;\frac{(a_4^2 - a_3^2a_1a_2)^2}{(a_3a_4)^2\,\overline{z}}$$

$$= \frac{1}{\frac{a_3a_4}{a_4^2 - a_3^2a_1a_2}\,\overline{z}} = \frac{1}{\overline{\overline{z}}}$$

Therefore $k(a_1,a_2,a_3,a_4,a_5) = k(a_1,a_2,a_3,a_4,\overline{\overline{z}})$

Claim: The invariants of β acting on $k(a_1,a_2,a_3,a_4,a_5)$ is

$$T = k(a_3,a_4,\ \bar{\bar{z}} + \frac{1}{\bar{\bar{z}}},\ a_1\bar{\bar{z}} + a_2\frac{1}{\bar{\bar{z}}},\ a_2\bar{\bar{z}} + a_1\frac{1}{\bar{\bar{z}}})$$

T is clearly fixed by β.

$[T(\bar{\bar{z}}):T] = 2.$

$a_3,a_4,\bar{\bar{z}} \in T(\bar{\bar{z}})$. One can solve for a_1 and a_2 since $\begin{vmatrix} \bar{\bar{z}} & \frac{1}{\bar{\bar{z}}} \\ \frac{1}{\bar{\bar{z}}} & \bar{\bar{z}} \end{vmatrix} \neq 0.$

Therefore $T(\bar{\bar{z}}) = k(a_1,a_2,a_3,a_4,a_5)$.

Since T is of codimension 2, it is the fixed field of β .

If $1 \to R_1 \to F_1 \to G \to 1$ is any other presentation of $G = Z_2 X Z_2$, then by a change of variables, one can assume F is free on x,y, and $z_1,\ldots,z_n$ and the map is x to (1,0), y to (0,1), and $z_i \to (0,0)$. It now follows that $R_1 = R \oplus Z[G]^n$. If t is a primitive element for $Qk[R]$ over $Qk[R]^G$, then $Qk[R_1]^G$ is formed by adding the varibles trace (z_i), trace (tz_i), trace (t^2z_i), and trace (t^3z_i).

This field is clearly invariant, if we adjoin t we obtain Q k[R]. One can then solve for the z_i since the Vandermonde determinent

$$\begin{vmatrix} 1 & 1 & 1 & 1 \\ t & \alpha(t) & \beta(t) & \alpha\beta(t) \\ t^2 & \alpha(t)^2 & \beta(t)^2 & \alpha\beta(t)^2 \\ t^3 & \alpha(t)^3 & \beta(t)^3 & \alpha\beta(t)^3 \end{vmatrix} \neq 0.$$

Since t satisfies a polynomial of degree 4, we are done.

§4. The Dihedral Groups

An extension L of k is called stably rational if $L(x_1,\ldots,x_n) = k(y_1,\ldots,y_m)$ for some indeterminates x_i and y_i. Clearly the conclusion of theorem 2.1 remains true if one only assumes $Qk[\overline{R}]^G$ is stably rational. (Apply Bloch's theorem twice.)

Let A be a G-lattice , that is A is a finitely generated torsion free abelian group which is a G-module. A is a permutation module if A has a basis permuted by G.

Suppose G also acts on k. If A is a G-lattice, then G acts in an obvious fashion on $Qk[A]$. One can ask when $Qk[A]^G$ is stably rational over k^G. The following theorem gives a complete answer:

Theorem 1. (Endo-Miyata [7], Voskresenskii [20]). Let G be a finite group and A a G-lattice. If G acts faithfully on the field k, then the following are equivalent.

(1) $Qk[A]^G$ is stably rational over k^G

(2) There is an exact sequence of G-lattices

$$1 \to A \to M \to N \to 1$$

where M and N are permutation modules.

We will apply this result to the dihedral group D_n of order 2n. Of course G acts trivially on k in the situation we are interested in. We solve the problem by writing $\overline{R} = A \oplus B$ such that A satisfies the hypothesis of the theorem, G acts faithfully on B, and $Qk[B]^G$ is purely transcendental over k.

We will consider only the standard presentation of D_n. That is $D_n = \langle x,y \mid x^n = 1,\ y^2 = 1,\ yxy^{-1} = x^{-1}\rangle$.

Theorem 2. If $1 \to R \to F \to D_n \to 1$ is the standard presentation, n is odd, char k is prime to n, and k contains a primitive nth root of unity, then $Qk[R]^{D_n}$ is stably rational over k.

As a corollary, we have

Theorem 3. If A is a crossed product with group D_n, n odd, char A prime to 2n, and the center of A contains a primitive nth root of unity, then A is similar to a product of cyclic algebras.

Let P be the permutation module given by the coset representation of the subgroup <y>. P has a basis $a_1,\ldots,a_n$ such that x permutes the a_i cyclically, $a_1^y = a_1$, $a_i^y = a_{n+2-i}$ for $i > 1$.

Lemma. If k contains a primitive nth root of unity, n is odd, and char k is prime to n, then $Qk[P]^G$ is purely transcendental over k.

Proof. Let ω be a primitive nth root of unity. Let

$b_i = \sum_{j=1}^{n} (\omega^{i-1})^{j-1} a_j$. Then $b_i^x = \omega^{n+1-i} b_i$. We then have $Qk(P) = k(b_1,\ldots,b$

since we can solve for the a_i using Cramer's rule.

$$k(b_1,\ldots,b_n)^x = k(b_1, b_n^n, \frac{b_2}{b_n^{n-1}}, \frac{b_{n-1}}{b_n^2}, \frac{b_3}{b_2^2}, \frac{b_{n-2}}{b_n^3}, \frac{b_4}{b_2^3}, \ldots)$$

since the right side is invariant under x and $k(b_1,\ldots,b_n)$ is obtained by adjoining b_n. Let $t = \frac{b_2}{b_n^{n-1}}$.

Claim: $k(b_1,\ldots,b_n)^{D_n} = k(b_1, b_n b_2, t + t^y, \frac{b_{n-1}}{b_n^2} + \frac{b_3}{b_2^2}, t\frac{b_{n-1}}{b_n^2} + t^y \frac{b_3}{b_2^3},$

$$\frac{b_{n-2}}{b_n^3} + \frac{b_4}{b_2^3}, t\frac{b_{n-2}}{b_n^3} + t^y \frac{b_4}{b_2^3}, \ldots)$$

is is of codimension 2 in $k(b_1,\dots,b_n)^x$ since trace (t) and norm (t) e in the right side. The right hand side is also clearly invariant.

<u>Proof of theorem 1</u>. A Schreier system is

$,x,x^2,\dots,x^{n-1},y,xy,x^2y,\dots,x^{n-1}y$. It follows that a free set of generators r $\bar{R}$ is

$$t = x^n,\ u_1 = y^2,\ u_2 = xy^2x^{-1},\ u_3 = x^2y^2x^{-2},\dots,u_n = x^{n-1}y^2x^{1-n}$$

$$v_1 = xyxy^{-1} \quad v_2 = x^2yxy^{-1}x^{-1},\dots,v_n = x^nyxy^{-1}x^{1-n} = yxy^{-1}x$$

fixes t and permutes the u_i's and v_i's cyclically.

We compute the action of y.

$$u_1^y = u_1,\quad u_2^y = u_1^{xy} = u_1^{yx^{-1}} = u_1^{x^{-1}} = u_n$$

$$u_3^y = u_{n-1},\quad u_4^y = u_{n-2},\dots ,$$

w $v_1^y = yxyxy^{-2} = (yxy^{-1}x)(x^{-1}yyx)y^{-2}$

$$= v_nu_nu_1^{-1} = u_1^{-1}u_nv_n$$

$$v_2^y = v_1^{xy} = v_1^{yx^{-1}} = (u_1^{-1}u_nv_n)^{x^{-1}} = u_n^{-1}u_{n-1}v_{n-1}$$

$$v_3^y = u_{n-1}^{-1}u_{n-2}v_{n-2},\dots,v_n^y = u_2^{-1}u_1v_1 .$$

lso $v_1\dots v_n = (yxy^{-1}x)(x^{n-1}yxy^{-1}x^{2-n})(x^{n-2}yxy^{-1}x^{3-n})\dots(x^2yxy^{-1}x^{-1})(xyxy^{-1})$

$$= yxy^{-1}x^nyx^{n-1}y^{-1} = y(xy^{-1}x^nyx^{-1})x^ny^{-1}$$

$$= y(yx^{-1}x^nxy^{-1})x^ny^{-1}$$ since xy^{-1} and yx^{-1} have the same action on $\bar{R}$

$= y^2x^ny^{-2}yx^ny^{-1}$

$= t^{y^2}t^y = t\ t^y$

Therefore $t^y = t^{-1}v_1\ldots v_n$.

Let $w_1 = u_{\frac{1+n}{2}}\, u^{-1}_{\frac{n-1}{2}}\, v_{\frac{1+n}{2}}$ and $w_{i+1} = w_i^{x^i}$

Then $w_1^y = w_1$ and the w_i are permuted cyclically by x.

Hence the action on the w_i is the same as on the u_i.

Clearly $t, u_1,\ldots,u_n, w_1,\ldots,w_n$ is a basis for $\overline{R}$.

Also $v_1\ldots v_n = w_1\ldots w_n$ so $t^y = t^{-1}\ w_1\ldots w_n$

Therefore $\overline{R} = \langle v_1,\ldots,v_n\rangle \oplus \langle t,w_1,\ldots,w_1\rangle$ as G-modules. Clearly $\langle v_1,\ldots,v_n\rangle$ is isomorphic to P in the lemma. Also D_n acts faithfully on P. Therefore if we can show that $\langle t,w_1,\ldots,w_n\rangle$ satisfies the hypothesis of theorem 1, the proof will be complete. This follows immediately from the

Claim: $\langle t,w_1,\ldots,w_n\rangle \oplus Z \cong P \oplus B$ where D_n acts on Z trivially, B is free on b_1 and b_2, x acts trivially on B, and y interchanges b_1 and b_2.

Define $f: P \oplus B \to \langle t,w_1,\ldots,w_n\rangle$ by $f(a_i) = (w_i,k)$,

$$f(b_1) = (t,\ell),\ f(b_2) = (t^{-1}w_1\ldots w_n,\ell).$$

f is clearly a G-map. If we compute the matrix of f relative to the ordered bases $a_1,\ldots,a_n,b_1,b_2$ and $w_1\ldots w_n,t,1$, we get

$$\begin{pmatrix} 1 & 0 & 0 & \ldots & 0 & 0 & k \\ 0 & 1 & 0 & \ldots & 0 & 0 & k \\ \vdots & & \vdots & & & & \vdots \\ 0 & 0 & 0 & \ldots & 1 & 0 & k \\ \cdots & & & & 0 & 1 & \ell \\ 1 & 1 & 1 & \ldots & 1 & -1 & \ell \end{pmatrix}$$

Adding the negative of the first n rows to the last and the n+1st row to the last we obtain

$$\begin{pmatrix} 1 & 0 & 0 & \ldots & 0 & 0 & k \\ 0 & 1 & 0 & \ldots & 0 & 0 & k \\ \cdot & \cdot & \cdot & \cdot & \cdot & \cdot & \cdot \\ 0 & 0 & 0 & \ldots & 1 & 0 & k \\ 0 & 0 & 0 & \ldots & 0 & 1 & \ell \\ 0 & 0 & 0 & \ldots & 0 & 0 & 2\ell-nk \end{pmatrix}$$

Since 2 and n are relatively prime, we can choose ℓ and k with $2\ell-nk = 1$. With these choices, f is an isomorphism.

§5. Questions

1. The basic question of the paper is whether the Brauer group is generate by cyclics. A weaker version is
2. Does every division algebra have an abelian splitting field? Both of these are probably false. Perhaps using the techniques of this paper, one could answer
3. If D has an abelian splitting field, is D similar to a product of cyclics?
4. Is theorem 3.2 true without the assumption of a primitive 4th root in the center? The techniques of this paper break down here.
5. An old question is whether a division algebra with involution is a product of quarternions. A positive solution is only know for degrees 2 and 4. Recently Tignol [19] proved that if the degree of D is 8, then $M_2(D)$ is a product of quarternions. It might be possible to prove that all division algebras with involution are similar to a product of quaternions if the general conjecture is false.
6. The method of proof for theorems 3.2 and 4.3 implies there is a fixed bound for the number of cyclics required. What is it? Could it be 2 in theorem 3.2?
7. A similar program can the carried out for the division ring of fraction of n×n generic matrices. If the center there is purely transcendental, then Bloch's theorem applies. Similar transfer techniques work. The hope of this paper is that our rings are easier to study then n×n generic matrices. For instance, to prove theorem 3.2, it would be

necessary to show the center of the division ring of fractions of 4×4 generic matrices is purely transcendental. The best result known is that the center of 3×3 generic matrices is purely transcendental [9].

Virginia Polytechnic Institute and State University
Blacksburg, Va. 24061

and

The Institute for Advanced Studies
The Hebrew University of Jerusalem, Israel

References

1. A.A. Albert, Structure of Algebras, Amer. Math. Soc. Coll. Pub. Vol. 24, Providence,Rhode Island, 1961.

2. S.A. Amitsur, On central division algebras, Israel J. of Math. 12(1972), 408-420.

3. S.A. Amitsur and D. Saltman, Generic abelian crossed products and p-algebras, J. of Algebra (to appear).

4. S. Bloch, Torsion algebraic cycles, K_2, and the Brauer group of function fields, Bull. A.M.S. 80(1974), 941-945.

5. P.M. Cohn, Algebra II, John Wiley, New York, 1977.

6. P. Deligne, Varieties unirationnelles non rationellos, Seminaire Bourbaki, Expose 402, Lecture Notes in Math., vol. 317, Springer-Verlag, New York, 1973.

7. S. Endo and T. Miyata, Invariants of finite abelian groups, J. Math. Soc. Japan, 25(1973), 7-26.

8. D.R. Farkas, Miscellany on Bieberbach group algebras, Pacific J. Math 59(1975), 427-435.

9. E. Formanek, The center of the ring of 3×3 matrices, Linear and Multilinear Algebra (to appear).

10. K.W. Gruenberg, Relation modules of finite groups, CBMS conference series, Vol. 25, American Math. Soc., Providence, Rhode Island.

1. K.W. Gruenberg, Cohomological topics in group theory, Lecture Notes in Math, Vol. 143, Springer-Verlag, New York, 1970.

2. M. Hall, The Theory of Groups, Macmillan, New York, 1959.

3. I.N. Herstein, Noncommutative Rings, John Wiley, New York, 1968.

4. P. Linnel, Zero divisors and idempotents in group rings, Math. Proc. Camb. Phil. Soc. 81(1977), 365-368.

5. J. Milnor, Introduction to Algebraic K-theory, Ann. of Math.Studies, no. 12, Princeton Univ. Press, Princeton, N.J., 1971.

6. J.P. Murre, Reduction of the proof of the non-rationality of a non-singular cubic threefold to a result of Mumford, Compositio Math. 27(1973), 63-82.

7. S. Rosset, Generic matrices, K_2, and unirational fields, Bull. A.M.S. 81(1975), 707-708.

8. S. Rosset, Abelian splitting of division algebras of prime degrees, Comment. Math. Helvetici 52(1977), 519-523.

9. J. Tignol, Sur les classes de similitude de corps a involution de degre 8, (to appear).

10. V.E. Voskresenskii, On the question of the structure of the subfield of invariants of a cyclic group of automorphisms of the field $Q(x_1,\ldots,x_n)$, (Russian). Izv. Akad. Nauk SSSR Ser. Mat. 34(1970), 366-375. English translation: Math. USSR-Izv. 4(1970), 371-380.

K-THEORY OF NOETHERIAN GROUP RINGS

J.T. STAFFORD[1]

Department of Mathematics, Brandeis University,[2]
Waltham, Mass. 02154.

The results described in this article are joint work with K.A. Brown and T.H. Lenagan. For reasons of space, we will only outline the proofs here, and for the full generalities, applications and proofs of these results, the reader is referred to "K-theory and stable structure of Noetherian group rings", to appear, by K.A. Brown, T.H. Lenagan and the present author. Hereafter this paper will be referred to as BLS .

[1] Supported by the British Science Research Council through a NATO Research Fellowship.

[2] Present address Gonville and Caius College, Cambridge, England.

A considerable body of research has been published on the -theory of group rings of finite groups, see for example [20]. We onsider here the corresponding problems and questions for the group rings f polycyclic by finite groups and show that many of the results do generalze to this case. In particular versions of the three basic stability heorems - Serre's Theorem (which says that modules of large rank have ree direct summands), the cancellation Theorem (which says that the omplementary direct summand is unique) and the Stable Range Theorem - do old in this case. See Theorems 4.5, 4.6 and 4.7 for exact details. his article is mainly concerned with the proofs and applications f these three theorems.

The proofs of the stability theorems are heavily dependent on the esults of [18] where they are given for a class of Noetherian rings called eakly ideal invariant. Thus to prove them for group rings it suffices to how weak ideal invariance, which we do in Sections 2 and 3. The rank of module used in [18] is defined in terms of homomorphisms from the module o the ring. However in the case of group rings of polycyclic by finite groups e are able to replace this by a rank defined in terms of various localisations f the group ring. In particular this notion of rank reduces to the familiar -rank used for modules over group rings of finite groups. See Section 4 or details.

The above results have all been concerned with arbitrary finitely enerated modules. However, if we specialise to projective modules over ntegral group rings, the definition of rank becomes particularly elementary; t is just the ratio of the uniform dimensions of the module and ring. his number is actually an integer and this has as an easy consequence the following result of Farkas and Snider [2]. If K is a field of characteristic zero and G is a torsion free polycyclic by finite group, then KG is a domain. We also give various conditions under which projectives become free. For example, if G is polycyclic by finite then there exists a normal subgroup H of finite index in G such that every finitely generated projective ZG-module is free as a ZH-module.

Other applications of these results deal with the structure of GL_n. For example if G is poly (infinite) cyclic then $GL_n(ZG)$ is finitely generated for large n.

2 WEAK IDEAL INVARIANCE AND STABLE STRUCTURE THEOREMS

Crucial to the results of this article are the results of [18] which prove the three stability theorems for a class of Noetherian rings known as weakly ideal invariant rings. In this section we present these results from [18] and prove various results about weak ideal invariance that will be useful when we turn to group rings.

The statements and proofs in [18] use the notion of Krull dimension of Rentschler and Gabriel and the reader is referred to [5] for the definition and basic properties of this dimension. Given a (right) module M over a ring R the <u>Krull dimension</u> of M is written $\text{kdim}_R M$ and the suffix will be dropped whenever no confusion can arise. An ideal T of a right Noetherian ring R is said to be <u>weakly ideal invariant</u> if, given any finitely generated right R-module M with $\text{kdim } M < \text{kdim } R/T$, then $\text{kdim } M \otimes T < \text{kdim } R/T$. Equivalently, T is weakly ideal invariant if, given a right ideal I with $\text{kdim } R/I < \text{kdim } R/T$, then $\text{kdim } T/IT < \text{kdim } R/T$. The ring is said to be weakly ideal invariant if every ideal is.

Finally we require a notion of rank. If M is a module over a ring R, then for a given integer ,

r-$rk(M) \geq s$ if, given any elements $\alpha_1,\ldots,\alpha_{s-1} \in M$ and P any prime of R, there exists $\alpha_s \in M$ and $\theta \in Hom(M,R)$ such that, for $1 \leq i \leq s-1$, $\theta(\alpha_i) \in P$ yet $\theta(\alpha_s) \in C(P)$. Here $C(P)$ denotes the elements of R that become regular in R/P.

2.1 SERRE's THEOREM [18 Theorem 2.1]. Let R be a right Noetherian weakly ideal invariant ring with kdim $R = n$. Suppose M is a left R-module with r-$rkM \geq n + 1$. Then $M \cong M' \oplus R$ for some module M'.

2.2 CANCELLATION THEOREM [18 Corollary 2.3]. Let R and M be as above and suppose that $M \oplus R \cong N \oplus R$ for some module N. Then $M \cong N$.

2.3 STABLE RANGE THEOREM [18 Corollary 2.4]. Let R be as above and suppose that $R = \Sigma_1^{m+1} a_i R$ for some $m \geq n+1$ and $a_i \in R$. Then there exist $f_i \in R$ such that $R = \Sigma_1^m (a_i + a_{m+1} f_i) R$.

Thus to prove versions of these theorems for various group rings, it suffices to show that the group rings in question are weakly ideal invariant. This we shall do in the next section. For the remainder of this section, we shall give various algebraic conditions that are sufficient to prove weak ideal invariance and will be useful later. It should be noted that it is easy to show that commutative rings and simple rings are weakly ideal invariant. Various other classes of Noetherian rings have also been so shown, including FBN rings, Asano orders and rings of Krull dimension one. See [BLS; 8; 9; and 17] for these and other examples. Indeed, there is no known example of a Noetherian ring that is not weakly ideal invariant, although it is easy to find right Noetherian rings that are not (see [9]).

We start by noting that a (non-trivial) maximality argument can be used to prove the following result.

LEMMA 2.4 Let R be a right Noetherian ring such that all prime ideals of R are weakly ideal invariant. Then so is R.

Proof. [BLS, Theorem 3.6]. □

Let I be an ideal of a ring R. Then I is said to be <u>right localizable</u> if $\mathcal{C}(I)$ satisfies the <u>right Ore</u> condition; that is, if $c \in \mathcal{C}(I)$ and $r \in R$, then there exists $c' \in \mathcal{C}(I)$ and $r' \in R$ so that $cr' = rc'$. I is said to be <u>localizable</u> if it is right and left localizable. There has been much research into the localizability of ideals of non-commutative rings, but they are of interest in the present circumstances because of the following:

PROPOSITION 2.5 [BLS, Corollary 3.8]. Let P be a right <u>localizable</u> prime ideal of a right Noetherian ring R. Then P is weakly ideal invariant

Proof. Let I be a right ideal of R such that $\text{kdim } R/I < \text{kdim } R/P$. Then there exists $c \in I \cap \mathcal{C}(P)$. So, if $x \in I \cap P$ there exists $r' \in R$ and $c' \in \mathcal{C}(P)$ such that $cr' = xc'$. As $x \in P$ and $c' \in \mathcal{C}(P)$, we have $r' \in P$ and $xc' = cr' \in IP$. But x was an arbitrary element of $I \cap P$. So $I\cap P/IP$ is a torsion R/P - module and $\text{kdim } I\cap P/IP < \text{kdim } R/P$. Hence

$$\text{kdim } P/IP = \max\{\text{kdim } P/I\cap P, \ \text{kdim } I\cap P/IP\} < \text{kdim } R/P .$$

Thus P is weakly ideal invariant. □

In particular a ring R has localizable prime ideals if it is AR.

An ideal T of R is said to be right AR if, given any right ideal there exists an integer n such that $IT \supset I T^n$. The ideal T is aid to be AR if it is right and left AR and R is AR if all ideals f R are AR). It is an open question whether a prime AR ideal P f a Noetherian ring is localizable and this question is related to the ossibility of P being weakly ideal invariant. For, it can be shown BLS, Theorem 3.11] that a prime AR ideal of a right Noetherian ring s weakly ideal invariant if and only if it is localizable. However f the ring is AR there is no problem.

OROLLARY 2.6 If R is a Noetherian AR ring then R is weakly ideal nvariant.

roof. By [16, Proposition 3.4] every prime of R is localizable. So he result follows from Proposition 2.5 and Lemma 2.4.□

Further results on the relationship between weak ideal invariance nd localizability are given in Section 3 of [BLS]. Corollary 2.6 is, owever, sufficient for our purposes.

WEAK IDEAL INVARIANCE OF GROUP RINGS

In this section we assemble various facts about the group rings hat we shall be considering, and show that these rings are weakly ideal nvariant of crucial importance are the recent results of Roseblade [13] nd we start with them.

Recall that a group G is called <u>polycyclic by finite</u> if there xists a chain,

$$1 = G_o \subset G_1 \subset \dots \subset G_n = G,$$

of subgroups of G such that each is normal in the next, and the factor groups are either finite or cyclic. It should be noted that the group ring RG of a polycyclic by finite group G over a commutative Noetherian ring R is Noetherian [7] and indeed these are the only known Noetherian group rings. Following Roseblade we say that a group G is orbitally sound if, given a subgroup H of G such that $|G: N_G(H)| < \infty$, then $|H^G: \mathrm{Core}_G(H)| < \infty$, where H^G is the normal closure of H and $\mathrm{Core}_G(H)$ is the intersection of the conjugates of H in G. A finite group H is said to be p-nilpotent if there exists a normal subgroup H' of H such that H/H' is a p-group and $p \nmid |H'|$. A polycyclic by finite group is p-nilpotent if all its finite images are p-nilpotent. The relevance to us of these two concepts comes from the following two results due to Roseblade.

PROPOSITION 3.1 Let G be a polycyclic by finite group. Then there exists a normal subgroup H of finite index in G such that H is poly (infinite) cyclic, orbitally sound and p-nilpotent.

Proof. Use [13, Theorem C 2] and [12, Lemma 11.2.16].

PROPOSITION 3.2 Let H be a polycyclic orbitally sound p-nilpotent group and k a field of characteristic p. Then kG is an AR-ring and hence weakly ideal invariant.

Proof. The first result is a consequence of Roseblade's work on the prime ideal structure of orbitally sound groups (see [13, Sect. 2.5]). The second result follows from this and Corollary 2.6. □

Of course, we want to show that if k is a field of positive characteristic and G is polycyclic by finite, then kG is weakly ideal invariant. Fortunately, this is a not too difficult consequence of the preceding results and we give an outline. The full proof is given in [BLS, Sect. 5]. Choose a normal subgroup H of finite index in G (by the above two propositions) such that kH is weakly ideal invariant. First, if P is a sub-bimodule of the free module $kH^{(r)}$ and M is a right kH-module with $\text{kdim}\, M < \text{kdim}\, kH^{(r)}/P$, then an easy induction on r shows that $\text{kdim}\, M \otimes P < \text{kdim}\, kH^{(r)}/P$. Now, if M is a finitely generated right kG-module, then by [14, Lemma 8] $\text{kdim}_{kG} M = \text{kdim}_{kH} M$. Note also that there exists an integer r such that $RG_{RH} \cong RH^{(r)}$. So given an ideal P of kG and a finitely generated kG-module M with $\text{kdim}\, M < \text{kdim}\, kG/P$, the above three comments can be combined to show that

$$\text{kdim}_{kG}\, M \otimes_{kG} P = \text{kdim}_{kH}\, M \otimes_{kG} P \leq \text{kdim}_{kH}\, M \otimes_{kH} P < \text{kdim}_{kH}\, kG/P$$

$$= \text{kdim}_{kG}\, kG/P\ .$$

Thus we have shown :

THEOREM 3.3 [BLS, Theorem 5.5]. If G is a polycyclic by finite group and k is a field of positive characteristic then kG is weakly ideal invariant.

If K is a field of characteristic zero then it is unknown whether KG is weakly ideal invariant - one trouble being that in general there are too few ideals in KG with the AR property. The best possible result at this stage is the following :

THEOREM 3.4 [BLS, Theorem 5.6]. Let K be a field of characteristic zero and G a nilpotent by finite group. Then KG is weakly ideal invariant.

Proof. Let H be a normal nilpotent subgroup of finite index in G. Then by [12, Theorems 11.2.8 and 11.3.12], KH is an AR ring and hence is weakly ideal invariant by Corollary 2.6. The argument given before Theorem 3.3 shows that KG is also

Fortunately, for certain rings of characteristic zero it is possible to show that the corresponding group ring is weakly ideal invariant .

THEOREM 3.5 Let R be a commutative domain that is finitely generated as a Z-algebra and let G be a polycyclic by finite group. Then RG is weakly ideal invariant.

This theorem is a special case of [BLS, Theorem 6.12] which proves the result for group rings over a somewhat larger class of commutative domains. The proof is considerably more difficult than that of Theorem 3.3 and depends heavily on the deeper results of [13] . Here we will just mention one crucial step in the proof, reminiscent of the Principal Ideal Theorem and of [14, Theorem F'], that may be of independent interest.

PROPOSITION 3.6 [BLS, Theorem 6.7]. Let R be a commutative domain, not a field, that is finitely generated as a Z-algebra and let G be a polycyclic by finite group. Suppose I is an ideal of RG such that RG/I is torsion-free as an R-module. Then there exists an ideal W of R such that

$$\text{kdim } RG/I = 1 + \text{kdim } RG/(I + WG).$$

4 THE RANK OF A MODULE

Since we know that various group rings are weakly ideal invariant, versions of the stable structure theorems hold for these rings. We shall present these versions at the end of this section, but we first want to improve them by giving a more intuitive notion of rank. If R is a commutative ring, then the usual concept of rank of an R-module M is given by the f-rank, defined as follows:

$$f - rk(M) \geq s \quad \text{if} \quad M_P \cong R_P^{(s)} \oplus M' \quad \text{for some module } M' = M'(P),$$

as P ranges through all maximal ideals of R.

It has been shown in [18], for a finitely generated module M over a commutative Noetherian ring R, that f-rkM = r-rkM. Now for the group rings under consideration, there do exist a large number of localizable ideals and it is natural to ask whether in this case we can replace r-rk by some analogue of f-rk. We answer this question affirmatively in this section. The first result that we give shows that there exist enough localizable ideals. If G is a group and R a ring we denote by $\underline{\underline{g}}$ the augmentation ideal of RG.

PROPOSITION 4.1 Let R be a field of positive characteristic or a commutative domain that is finitely generated as a Z-algebra and let G be a polycyclic by finite group. Suppose P is a maximal ideal of RG. Then:

i) There exists a normal subgroup H of finite index in G such that $\mathcal{C} = \mathcal{C}_{RH}(P \cap RH)$ is a right divisor set of regular elements of RG.

ii) If R is a finitely generated Z-algebra, then in i) H can be chosen such that $P \cap RH = P_o RH + \underline{\underline{h}}$ for some maximal ideal P_o of R.

iii) Suppose H is chosen as in i). Then $RG_{\mathcal{C}}$ is a semilocal ring containing $(P \cap RH)RG$ in its Jacobson radical.

Proof. For full details see [BLS Proposition 7.5]. We shall give the proof when R is a field of characteristic p. By Proposition 3.1, we may choose a normal subgroup H of finite index in G such that H is poly (infinite) cyclic, orbitally sound and p-nilpotent. By Proposition 3.2, RH is an AR ring and so by [16 Proposition 3.4] we can localise at every semiprime ideal of RH. But by [13, Lemma 5] $P\cap RH$ is a semiprime ideal of RH. So $\mathcal{C} = \mathcal{C}_{RH}(P\cap RH)$ is an Ore set in RH. As H is poly (infinite) cyclic, RH is a domain so $\mathcal{C}$ consists of regular elements. Since G/H is finite it is easy to check that $\mathcal{C}$ satisfies the required conditions for i). Now $T = RH_{\mathcal{C}}/(P\cap RH)_{\mathcal{C}}$ is an artinian ring. Since $RG_{\mathcal{C}}/PRG_{\mathcal{C}}$ is a finitely generated module over T, it is itself artinian and part iii) follows easily □

REMARK 4.2 It should be noted that the localisations described in Proposition 4.1 are in fact localizations of RG at semiprime ideals of RG. For, choose H by Proposition 4.1 i) and set $I = P\cap RH$ and $\mathcal{C} = \mathcal{C}_{RH}(I)$. Using Proposition 4.1 iii), it is easily checked that $RG_{\mathcal{C}} = RG_N$ where N is the nilradical of IG.

It is possible at this stage to define a variant of f-rk using the localizations provided by the last proposition. However, it is appropriate to make the definitiion in a more general setting that has the merit of being a proper generalization of the f-rk of a commutative ring. We would like to thank B. Mueller for suggesting the possibility of this more general setting.

Following [11], we say that a set $\{P_1,\ldots,P_n\}$ of prime ideals of a Noetherian ring S is a <u>clan</u> if i) $N = \cap P_i$ is a localizable ideal of S; ii) the ideal N_N is an AR ideal of S_N; and iii) no proper subset of $\{P_1,\ldots,P_n\}$ satisfies conditions i) and ii). If $\{P = P_1,P_2,\ldots,P_n\}$ is a clan we say that P <u>belongs to a clan</u> and we call N the <u>chieftain</u> of P. By [11, Theorem 5], a prime ideal P belongs to at most one clan. Further if P is maximal and every maximal ideal belongs to a <u>clan</u> then each P_i for $2 \le i \le n$ is itself maximal.

COROLLARY 4.3 Let R be a field of positive characteristic or a commutative domain that is finitely generated as a Z-algebra, and let G be a polycyclic by finite group. Then every maximal ideal of RG belongs to a clan.

We can now define our variant of f-rk in this setting. Suppose S is a Noetherian ring in which every maximal ideal belongs to a clan. For each maximal ideal P of S, fix a semimaximal localizable ideal $T \subseteq P$. We say that T is <u>associated</u> to P and write $S_T = S_{\tilde{P}}$. For an S-module M and integer s define

$$\text{q-rk } M \geq s \quad \text{if} \quad M_{\tilde{P}} = S_{\tilde{P}}^{(s)} \oplus M' \quad \text{for some module } M' = M'(P) \text{ as } P \text{ ranges through the maximal ideals of } S.$$

Of course it is by no means clear that q-rk is well defined in the sense of being independent of the choice of the associated ideals. However this follows immediately from the next result.

PROPOSITION 4.4 Let S be a Noetherian ring such that every maximal ideal of S belongs to a clan. Let M be a finitely generated S-module. Then $\text{q - rk}(M) = \text{r-rk}(M)$.

Proof. This can be proved by making the appropriate changes in the proof of [18, Proposition 2.6]. See [BLS, Theorem 7.9] for full details. □

The reason for giving the definition of q-rk in this more general setting is that in differing situations, different choices of the associated ideals seem to be most convenient. For the results of this article, the localizations given by Proposition 4.1 are the most useful. In general, there is a unique canonical choice of the ideal T associated to a given prime P of a ring S - the chieftain of P. It is this choice of T

that yields the f-rk when S is commutative. If S is the group ring of a finite group G over a commutative Noetherian ring R then the choice of T that gives the usual definition of rank is the nilradical of $(P\cap R)S$. See, for example,[20]. Note that this is not in general equal to the chieftain of P.

We end the section by restating the three basic stability theorems for group rings. These follow from the results of this and the previous two sections, once we recall that if G is a polycyclic by finite group and R is a commutative Noetherian ring, then $\text{kdim } RG = \text{kdim } R + h(G)$ where $h(G)$ denotes the Hirsch number of G (see [15]).

THEOREM 4.5 Let R be a field of positive charcteristic or a commutative domain that is finitely generated as a Z-algebra with $\text{kdim } R = t < \infty$. Let G be a polycyclic by finite group and suppose that M is a finitely generated RG-module with $q\text{-rk}(M) \geq t + h(G) + 1$. Then $M \cong RG \oplus M'$ for some RG-module M'.

THEOREM 4.6 Let R, G and M be as above and suppose for some module N that $M \oplus RG \cong N \oplus RG$. Then $M \cong N$.

THEOREM 4.7 Let R and G be as above. Suppose that $RG = \sum_1^{m+1} a_i RG$ for some $a_i \in RG$ and $m \geq h(G) + t + 1$. Then there exist elements $f_i \in RG$ such that $RG = \sum_1^m (a_i + a_{m+1} f_i) RG$.

Using Theorem 3.4, similar results can be given if R is a field of characteristic zero and G is nilpotent by finite. However, using Theorem 4.7, some information can also be obtained for an arbitrary polycyclic by finite group G.

COROLLARY 4.8 Let K be a field of finite transcendence degree n over Q and G be a polycyclic by finite group. Suppose that $KG = \sum_1^{m+1} a_i KG$ for some $a_i \in KG$ and $m \geq n + h(G) + 2$. Then there exist $f_i \in KG$ such that $KG = \sum_1^{m} (a_i + a_{m+1} f_i) KG$.

Proof. Suppose $1 = \sum_1^{m+1} a_i b_i$ for some $b_i \in KG$. Then there exists a finitely generated Z-algebra $R \subseteq K$ such that each a_i and $b_i \in RG$. Now apply Theorem 4.7, noting that kdim $R \leq n+1$. □

We conjecture that Corollary 4.8 holds whenever $m \geq h(G) + 1$, and with a little extra effort, this can be proved if K is an algebraic extension of Q (see [BLS, Theorem 9.2]). However, the naive approach used to prove Corollary 4.8 cannot, it seems, be used to obtain versions of Theorems 4.5 or 4.6 for group rings of arbitrary polycyclic by finite groups over fields of characteristic zero.

5 PROJECTIVE MODULES AND APPLICATIONS

In this section, we give various applications of the Theorems 4.5, 4.6 and 4.7 . We will be concerned primarily with the structure of projective modules and will give various conditions under which projectives become free. For example, projective modules over the integral group ring become free when we pass to certain localizations or consider them as modules over a suitable subring of finite index. However these results also provide some interesting information in other directions, giving for example a new proof of the Farkas-Snider result that KG is a domain whenever K is a field of characteristic zero and G is a torsion-free polycyclic by finite group.

We begin by considering large projective modules. A ring S is said to be <u>stably free</u> if every finitely generated projective S-module is stably free; that is, if M is a finitely generated projective S-module there exists integers r and s such that $M \oplus S^{(r)} \cong S^{(s)}$.

We denote the (Goldie) uniform dimension of module M by udim(M).

We first show that stably free group rings turn up with reasonable frequency.

PROPOSITION 5.1 [BLS, Corollary 8.2]. Let R be a field of positive characteristic or the integers. Suppose G is either poly(infinite) cyclic or torsion-free supersoluble. Then ;

a) RG is stably free.

b) Thus, if M is a finitely generated projective RG-module, there exist integers r and s such that $M \oplus RG^{(r)} \cong RG^{(s)}$. In this case q-rkM = s-r = udim M and M is free if $s-r \geq 1 + h(G) + \text{kdim}\, R$.

Proof. RG is stably free by [4, Theorems 25 and 27] and [10 Theorem B] respectively. Part b) now follows easily from Theorem 4.6. □

Other cases of stably free group rings are given in [BLS, Corollary 8.2 and 9.3]. Actually it is unclear whether there exist any non-free finitely generated projective modules over the rings considered in Proposition 5.1. Indeed, if G is free abelian then all projective RG-modules are free [21]. In this direction the proposition has the following amusing corollary.

COROLLARY 5.2 [BLS, Corollary 8.3]. Let R be a field of positive characteristic or the integers and let G be a polycyclic by finite group. Then there exists a normal subgroup H of finite index in G such that every finitely generated projective RG-module is free as an RH-module.

Proof. By [12 Lemmas 10.2.5 and 10.2.11] there exists a normal poly(infinite)cyclic subgroup H of finite index in G such that

$$n = (\text{kdim}\, R + h(G))\, \text{udim}\, RG < |G:H| < \infty .$$

Now $RG_{RH} \cong RH^{(n)}$ and so $\mathrm{udim}_{RH} M > \mathrm{kdim}\, R + h(G)$. Since RH is a stably free domain, $\text{q-rk}_{RH} M = \mathrm{udim}_{RH} M$ and by Proposition 5.1, M_{RH} is free. □

We now turn to the localizations of the integral group ring ZG given by Proposition 4.1. If G is finite one has as a particular case of these localizations, the rings $Z_p G$, where p runs through the primes of Z. In this case, it is known that given any finitely generated projective ZG-module M, M_p is free [20, Theorems 4.2 and 2.21]. This fact forms a crucial part of the proof of the next result.

THEOREM 5.3 Let R be a commutative Noetherian domain in which no rational prime is a unit and let G be a polycyclic by finite group. Let P be an ideal of R such that R/P is a field of characteristic q > 0. Choose a normal subgroup H of finite index in G such that $I = PG + \underline{\underline{h}}G$ is a localizable ideal of RG. Then, if M is a finitely generated projective RG-module, M_I is free.

Proof. We given an outline of the proof for R = Z. For full details see [BLS, Theorem 8.5]. Note that by Proposition 4.1, there exists a localizable ideal I of the required form. Now $M/M\underline{\underline{h}}$ is a finitely generated projective module over $ZG/\underline{\underline{h}}G \cong Z(G/H)$. Thus by [20, Theorems 4.2 and 2.21], $(M/M\underline{\underline{h}})_p$ is a free $Z_p(G/H)$-module, say isomorphic to $Z_p(G/H)^{(r)}$. Since $(\underline{\underline{h}}G + PG)_I \subseteq I_I \subseteq J(ZG_I)$, this pulls back to an isomorphism $M_I \cong (ZG_I)^{(r)}$, as required. □

One consequence of the theorem is that, for the rings under consideration, we have a very simple notion of rank.

COROLLARY 5.4 Let R, G and M be as in the theorem, and suppose R is finitely generated as a Z-algebra. Then RG has an artinian full quotient ring F and $M \otimes_{RG} F \cong F^{(r)}$ where

$$r = q - \mathrm{rk}(M) = \mathrm{udim}\, M/\mathrm{udim}\, RG .$$

Proof. For each maximal ideal P of RG choose an associated ideal T by Proposition 4.1 ii) and Remark 4.2. Then by Theorem 5.3 $M_{\tilde{P}}$ is free. Since by Proposition 4.1, $C(T)$ consists of regular elements of RG, F is also the full quotient ring of $RG_{\tilde{P}} = S$. So, for some integer r,

$$M \otimes_{RG} F \cong M_{\tilde{P}} \otimes_S F \cong S^{(r)} \otimes F \cong F^{(r)},$$

and the result follows. □

It should be noted that, in the notation of Theorem 5.3, we have <u>not</u> shown that all finitely generated projective RG_I-modules are free. For example, let $R = Z$ and $G = \langle x \rangle$ be a group of finite order n. Put $I = p\, ZG$ for some prime p of Z that is coprime to n. Then $1/n \sum x^i$ is a nontrivial idempotent of ZG_I and hence generates a non-free projective module. Similarly, by [6, Corollary 3.13 ii)], not all projective ZG - modules are free when G is a finite group.

As a second corollary to Theorem 5.3, we move in a different direction and give a new proof of the recent result of Farkas and Snider [2] on the zero divisor question.

COROLLARY 5.5 [2] Let G be a torsion-free polycyclic by finite group and K a field of characteristic zero. Then KG is a domain.

Proof. [BLS, Theorem 8.8]. It clearly suffices to prove the result when K is a finitely generated field extension of Q. In this case, it is possible to construct a Dedekind domain R which is an order in K such that no prime of Z is a unit in R . It is enough to show that RG is a domain. Now since KG is semiprime [12 Theorem 4.2.12], RG is at least semiprime. Given any finitely generated projective RG-module M then, by Corollary 5.4, udim RG divides udim M. But, since R is hereditary, RG has finite global dimension by [4, Theorem 25 b)] and [22, Theorem 9.2]. It follows from these last two comments, using a standard argument due to R. Walker (see for example [12, Theorem 10.4.13]), that udim RG divides udim N for any finitely generated torsion-free RG-module N. Since RG is semiprime, this is only possible if RG is a domain. □

If it were possible to obtain a version of Theorem 5.3 for group algebras of torsion-free polycyclic by finite groups over fields of positive characteristic, then the argument used in Corollary 5.5 would also prove the zero divisor question for this case. It is actually possible to do this for certain groups, although the proof depends upon the fact that the zero divisor question has been solved in these cases. See [BLS, Theorem 8.9] for details.

Given the considerable body of research on K-theory it is possible to give various fairly easy consequences of Theorems 4.5, 4.6 and 4.7 and we end by mentioning one such example. For further examples the reader is referred to [BLS, Sect. 8 and 10]. We consider the general linear group and give some conditions for its finite generation.

If A is a ring and m an integer let $E_m(A)$ denote the subgroup of $GL_m(A)$ generated by the elementary matrices

$$\{I_m + ae_{ij} : \quad i \neq j \text{ and } a \in A\}.$$

Given a group H we denote by H' the derived subgroup of H.

THEOREM 5.6. Let G be a poly (infinite)cyclic group or a finitely generated torsion-free abelian by finite group. Then for $m \geq h(G) + 3$ we have

$$GL_m(ZG) / GL_m(ZG)' \cong GL_m(ZG) / E_m(ZG) \cong K_1(ZG) \cong C_2 \oplus G/G' .$$

In particular $GL_m(ZG)$ is finitely generated for $m \geq h(G) + 3$.

Proof. The first two isomorphisms come from Theorem 4.7 and [23, Theorem 3.2]. The third comes from [4, Theorem 29] if G is poly (infinite cyclic), and from [3, Theorem 3.1] if G is abelian by finite. Finally, $GL_m(ZG)$ is finitely generated since, for $m \geq 3$, $E_m(A)$ is finitely generated wherever A is a finitely generated ring [1, V, Corollary 1.3] . □

Similar results hold for various other group rings as can be seen in [BLS, Theorems 10.2 and 10.3] .

REFERENCES

1. H. BASS, Algebraic K-theory, Benjamin, New York, 1968.

2. D.R. FARKAS and R.L. SNIDER, K_o and Noetherian group rings, J. Algebra 42 (1976), 192-198.

3. F.T. FARRELL and W.C. HSIANG, The topological - Euclidean space form problem, Inv. Math 45 (1978), 181-192.

4. F.T. FARRELL and W.C. HSIANG, A formula for $K_1 R_\alpha[T]$, Proc. Symp. Pure Math. vol. 17, (1970), 192-198.

5. R. GORDON and J.C. ROBSON, Krull dimension, Mem. Amer. Math. Soc. 133, (1973).

6. K.W. GRUENBERG, Relation modules of finite groups, Regional Conference Series in Math. No 25, Amer. Math. Soc., 1976.

7. P. HALL, Finiteness conditions for solvable groups, Proc. London Math. Soc. 4(1954), 419-436.

8. G. KRAUSE, T.H. LENAGAN and J.T. STAFFORD, Ideal invariance and Artinian quotient rings, J. Algebra, to appear.

9. T.H. LENAGAN, Noetherian rings with Krull dimension one, J. London Math. Soc. 15 (1977), 41-47.

10. P.A. LINNELL, Zero divisors and idemptotents in group rings, Proc. Camb. Phil. Soc. 81 (1977), No. 3, 365-368.

11. B.J. MUELLER, Localisation in non-commutative Noetherian rings, Can. J. Math. 28 (1976), 600-610.

12. D.S. PASSMAN, The algebraic structure of infinite group rings, Interscience, 1977.

13. J. E. ROSEBLADE, Prime ideals in group rings of polycyclic groups, Proc. London Math. Soc., 36 (1978), 385-447.

14. D. SEGAL, The residual simplicity of certain modules, Proc. London Math. Soc. 34 (1977), 327-353.

15. P.F. SMITH, On the dimension of group rings, Proc. London Math. Soc. 25 (1972), 288-302; Corrigendum, ibid. 27 (1973), 766-768.

16. P.F. SMITH, Localisation and the AR property, Proc. London Math. Soc. 22 (1971), 39-68.

17. J.T. STAFFORD, Stable structure of noncommutative Noetherian rings, J. Algebra 47 (1977), 244-267.

18. J.T. STAFFORD, Stable structure of noncommutative Noetherian rings II, J. Algebra, 52 (1978) 218-235.

19. R.G. SWAN, Algebraic K-theory, Lecture Notes in Math. No 76, Springer-Verlag, Berlin / New York, 1968.

20. R.G. SWAN, K-theory of finite groups and orders, Lecture Notes in Math. No 149, Springer- Verlag, Berlin / New York, 1970.

21. R.G. SWAN, Projective modules over Laurent polynomial rings, Trans. Amer. Math. Soc. 237 (1978), 111-121.

22. R.G. SWAN, Groups of cohomological dimension one, J. Algebra 12 (1969), 585-601.

23. L. N. VASERSTEIN, On the stabilization of the general linear group over a ring, Mat. Sb. 79 (121) (1969), 405-424; translated as Math. USSR Sb. 8 (1969), 383-400.

THE CANCELLATION PROBLEM FOR PROJECTIVE MODULES AND RELATED TOPICS*

by

A. A. Suslin

Introduction

The classification of finitely generated projective modules over a ring A can be, as a rule, divided into two rather different parts. The first, the classification up to stable isomorphism, is equivalent to the study of the Grothendieck group $K_0(A)$ and is traditional for algebraic K-theory. The second part, the cancellation problem, is studied much less than the first one. Most of the results in this field obtained up to 1972 can be found in Bass' talk [2]. I'll speak today about several new results in this direction obtained during the last few years. I'll begin with some definitions and general remarks.

Two finitely generated projective A-modules P and P' are called stably isomorphic if $P \oplus A^n \cong P' \oplus A^n$ for some n. We shall say that P satisfies the cancellation condition if any P' stably isomorphic to P is really isomorphic to P. What does this mean in other terms? Suppose that $P \oplus A \cong P' \oplus A$ and choose some isomorphism $\lambda : P' \oplus A \to P \oplus A$. The image $u = \lambda(0 \oplus 1) \in P \oplus A$ under λ of the element $0 \oplus 1 \in P' \oplus A$ is a unimodular element in $P \oplus A$ (i.e., there exists $\varphi : P \oplus A \to A$ with $\varphi(u) = 1$) and the module P' is isomorphic to $P \oplus A / A \cdot u$. On the other hand, if $u \in P \oplus A$ is a unimodular element and if we put $P' = P \oplus A / A \cdot u$, then $P \oplus A \cong P' \oplus A$. So we see

If P satisfies the cancellation condition, then $\mathrm{Aut}(P \oplus A)$ acts transitively on the unimodular elements in $P \oplus A$.

The converse is also true if one supposes in addition that $P \oplus A$ satisfies the cancellation condition.

---

*This is an expanded version of an invited address which was given on August 21, 1978, at the International Congress of Mathematicians in Helsinki, Finland.

The classical and quite easy cancellation theorem is the following: Modules of rank one always satisfy the cancellation condition. The proof is so simple that I'll remind you of it. Suppose $P \oplus A^n \cong P' \oplus A^n$, where rank $P =$ rank $P' = 1$, and take the (n+1)st exterior power. This yields $P \cong \bigwedge^{n+1}(P \oplus A^n) \cong \bigwedge^{n+1}(P' \oplus A^n) \cong P'$.

The other well-known cancellation theorem is a theorem of Bass [1] which states that modules of large enough rank satisfy the cancellation condition.

THEOREM 1 (Bass). If A is a commutative noetherian ring and P is a finitely generated projective module with rank $P \geq 1 + \dim \operatorname{Max} A$, then P satisfies the cancellation condition.

If one deals with the category of all commutative rings, then the inequality in Bass' theorem is the best possible; however for some special rings one can prove stronger cancellation theorems, and that will be one of our aims below.

I. Polynomial rings.

The cancellation problem for projective modules over polynomial rings is closely connected with the well-known Serre problem on freeness of projective modules over a polynomial ring over a field. In fact, it was proved by Serre [16] in 1958 that if P is a projective module over $k[X_1,\ldots,X_n]$, then P is stably isomorphic to a free module, so Serre's problem was equivalent to whether or not free $k[X_1,\ldots,X_n]$-modules satisfy the cancellation condition.

The main results in this direction were obtained in 1976 independently by D. Quillen [14] and myself [17]. There have been several talks on this theme since 1976 (see [7]) so I'll restrict myself to the formulation of the main theorem.

THEOREM 2. Let B be a commutative noetherian ring and $A = B[X_1,\ldots,X_n]$ a polynomial ring over B. If P is a finitely generated projective A-module of rank $\geq 1 + \dim B$, then:

.) *If* P *is extended from* B, *then* P *satisfies the cancellation condition.*

:) *If* B *is regular, then* P *is extended from* B.

This result was developed and generalized in different directions y several authors; there are also analogues of this theorem for ymplectic and quadratic modules, for the functor K_1, and so on. 'll mention here only two results: Richard Swan [24] has shown that heorem 2 is valid not only for a polynomial ring, but also for a aurent polynomial ring $B[X_1,\dots,X_k,X_{k+1}^{\pm 1},\dots,X_n^{\pm 1}]$. I have proved the ollowing noncommutative version of Theorem 2.

HEOREM 3. *Suppose that* B *is a finite module over its center* B_0 *hich is noetherian. If* P *is a finitely generated projective* $[X_1,\dots,X_k,X_{k+1}^{\pm 1},\dots,X_n^{\pm 1}]$*-module of rank* $>\max(\dim B_0,1)$, *then the ame conclusions as in Theorem 2 are valid.*

I want to mention that the new condition rank $P>1$, which ppears in Theorem 3 is essential, since Ojanguren and Sridharan [13] ave shown that for any noncommutative field B there exist finitely enerated projective $B[X_1,X_2]$ - modules of rank one which are not free nd hence not extended from B.

One of the main unsolved questions in the theory of projective odules over polynomial rings is the following one which was raised y Bass: Suppose A is a commutative regular ring; is it true that initely generated projective $A[X]$ - modules are extended from A ?

In view of Quillen's localization principle this question is quivalent to the following: Suppose A is a commutative regular ocal ring; is it true that finitely generated projective $A[X]$ - odules are free?

For two-dimensional rings the answer to the Bass-Quillen question s positive by the Horrocks-Murthy theorem ([11]). Furthermore, it may be shown that if the answer to the Bass-Quillen question is positive for all regular rings of dimension less than or equal to d, then more generally, for any regular ring A of dimension less than or equal to d all finitely generated projective $A[X_1,\ldots,X_n]$-modules are extended from A. In particular, finitely generated projective $A[X_1,\ldots,X_n]$-modules are extended from A if $\dim A \leq 2$. Mohan Kumar [10] and, independently, Lindel and Lütkebohmert [8], gave a positive answer to the Bass-Quillen question for rings of power series:

THEOREM 4. *If $A = k[[T_1,\ldots,T_d]]$ is a ring of formal power series over a field k, then any finitely generated projective $A[X_1,\ldots,X_n]$-module is free.*

Recently I've proved that the answer to the Bass-Quillen question is positive for certain three dimensional rings:

THEOREM 5. *If A is the coordinate ring of a smooth affine variety of dimension d over a field k and P is a finitely generated projective $A[X_1,\ldots,X_n]$-module, then P is extended from A in each of the following cases:*

1) $\operatorname{rank} P \geq d$,
2) $d \leq 3$ *and* $\operatorname{char} k \neq 2$.

The proof is based on the following idea. For any local ring A there exists a correspondence between stably free $A[X]$-modules and nilpotent endomorphisms of finitely generated A-modules, of the following type: To any pair (M,α), where M is a finitely generated A-module and α is a nilpotent endomorphism of M one can associate a stably free $A[X]$-module $P(M,\alpha)$ which is defined up to "near isomorphism" (cf. [25]), and every stably free $A[X]$-module P can be obtained in this manner. Moreover, if $\operatorname{rank} P = r$, then there exists (M,α) such that $P \cong P(M,\alpha)$ and $\dim M$ $(=\dim(A/\operatorname{Ann} M)) \leq \dim A - r$. Now if $\operatorname{rank} P = \dim A$, then the previous remark shows hat $P \cong P(M,\alpha)$, where M is a module of finite length.

In the situation of Theorem 5 suppose that $n=1$ and let μ be ny maximal ideal of A. By Quillen's localization principle it is nough to show that P_μ is a free $A_\mu[X]$-module. Since A_μ is regu-ar, P_μ is stably free and hence has the form $P_\mu = P(M,\alpha)$, where is an A_μ-module of finite length. It may be shown, using moothness, that A_μ contains a local subring B of the form $[X_1,\ldots,X_d]_\nu$ such that $\hat{B} = \hat{A}_\mu$. Since M is of finite length, it s defined over B; hence P_μ is defined over $B[X]$ and now it emains to use known results about projective modules over a poly-omial ring over a field.

I. Affine algebras over a field.

If A is an affine algebra over a field k, then $\dim \operatorname{Max} A =$ im A and Bass' theorem reads as follows: If rank $P \geq 1 + \dim A$, then satisfies the cancellation condition.

It is well-known that for affine algebras over the field $\mathbb{R}$ ass' theorem can not be strengthened: For any $n > 1$ there exist -dimensional affine $\mathbb{R}$-algebras A such that the A-module A^n does ot satisfy the cancellation condition. Such examples have topo-ogical origin and were first constructed by R. Swan in his paper 22]. Further information about connections between the topological nd algebraic situations can be found in his recent paper [23].

However in the case of affine algebras over algebraically closed ields the situation is different (see [18]):

HEOREM 6. Suppose A is an affine algebra over an algebraically losed field k and P is a finitely generated projective A-module f rank greater than or equal to $\dim A$. Then P satisfies the can-ellation condition.

I don't think that the bound in Theorem 6 is the best possible. I suppose that the correct bound would be $\operatorname{rank} P \geq (1+\dim A)/2$; however, this seems to be a rather difficult problem.

In the case of two-dimensional algebras, using Theorem 6 and the fact that modules of rank one always satisfy the cancellation condition, we obtain the theorem of Murthy-Swan [12]: <u>If A is a two-dimensional affine algebra over an algebraically closed field, then finitely generated projective modules satisfy the cancellation condition.</u>

Theorem 6 is closely connected to the following curious result of [19] on unimodular rows: If $v = (a_0, \ldots, a_r)$ is a unimodular row over a commutative ring A and $n_0, \ldots, n_r$ are natural numbers, whose product is divisible by $r!$, then the unimodular row $(a_0^{n_0}, \ldots, a_r^{n_r})$ can be completed to an invertible matrix. In the case $r = 2$, this result was independently proved by Swan and Towber [25], and they have also shown that the divisibility condition is necessary for its validity.

Let's accept this result and prove that A^n $(n = \dim A)$ satisfies the cancellation condition, where A is as in Theorem 6. Since A^{n+1} satisfies the cancellation condition by Bass' theorem, it is enough to show that $GL_{n+1}(A)$ acts transitively on the set $Um_{n+1}(A)$ of unimodular rows, or in other words to show that any unimodular row can be completed to an invertible matrix. We can restrict ourselves to the case of reduced algebras. If $v = (a_0, \ldots, a_n)$ is a unimodular row, then by adding some multiples of a_0 to $a_1, \ldots, a_n$ to bring these elements to general position, we can suppose that

$\dim A/(a_1, \ldots, a_n) = 0$. Moreover, by Bertini's theorem we can suppose that $A/(a_1, \ldots, a_n)$ is reduced. Since k is algebraically closed one has $A/(a_1, \ldots, a_n) \cong k \times \ldots \times k$. Using once more the fact that k is algebraically closed, we see that $a_0 \equiv b_0^{n!} \bmod (a_1, \ldots, a_n)$ for some b_0, and hence the row v can be transformed with the help of elementary transformations to $(b_0^{n!}, a_1, \ldots, a_n)$ and hence can be completed to an invertible matrix.

In view of Theorem 6 it is natural to ask for which fields k projective modules over affine k-algebras satisfy a strengthened cancellation theorem. It turns out that the answer depends on arithmetical properties of the field.

I'll discuss here the case of two-dimensional algebras, which is more or less well understood. As stated above, if A is such an algebra, then A^2 satisfies the cancellation condition if and only if the group $SL_3(A)$ acts transitively on the set $Um_3(A)$. A beautiful theorem of Vaserstein states that the set of orbits $Um_3(A)/SL_3(A)$ has a canonical abelian group structure (see [28]). More precisely, suppose A is any commutative ring and $v=(a_1,a_2,a_3)$ is a unimodular row of length three over A. Choose b_1,b_2,b_3 such that $\Sigma\, a_i b_i = 1$ and consider the matrix

$$\begin{pmatrix} 0 & a_1 & a_2 & a_3 \\ -a_1 & 0 & b_3 & -b_2 \\ -a_2 & -b_3 & 0 & b_1 \\ -a_3 & b_2 & -b_1 & 0 \end{pmatrix}$$

This matrix is alternating and unimodular, so it endows the module A^4 with a symplectic structure. Up to isomorphism, this symplectic module does not depend on the choice of the b's and does not change when the row v is replaced by $v\cdot\alpha$ $(\alpha \in SL_3(A))$. Thus we obtain a canonical mapping $Um_3(A)/SL_3(A) \to V(A) = \ker(K_0Sp(A) \to K_0(A))$, given by the formula

$$v \mapsto \left[\left(A^4, \begin{pmatrix} 0 & a_1 & a_2 & a_3 \\ -a_1 & 0 & b_3 & -b_2 \\ -a_2 & -b_3 & 0 & b_1 \\ -a_3 & b_2 & -b_1 & 0 \end{pmatrix}\right)\right] - \left[\left(A^4, \begin{pmatrix} 0 & 1 & 0 & 0 \\ -1 & 0 & 0 & 0 \\ 0 & 0 & 0 & 1 \\ 0 & 0 & -1 & 0 \end{pmatrix}\right)\right]$$

and the theorem of Vaserstein states that under certain conditions, in particular for two-dimensional rings, this mapping is a bijection.

Let k be an arbitrary field and denote by $A = A_k$ the two-dimensional affine k-algebra $k[X,Y,Z]/(X^2-X)(Y^2-Y)(Z^2-Z)$. Using excision theorems and some other methods, it may be shown that $V(A)$ equals $G(k) = \ker(K_2Sp(k) \to K_2(k))$. Let $W(k)$ denote the Witt ring of quadratic forms over k and $I(k)$ its maximal ideal consisting of even dimensional quadratic forms. The theorem of Matsumoto gives a description of $K_2(k)$ and $K_2Sp(k)$ in terms of generators and relations. From this description it follows that there exists a canonical epimorphism $K_2Sp(k) \to I^2(k)$ and the image of the group $G(k)$ under this epimorphism is just $I^3(k)$. (It seems rather probable that the epimorphism $G(k) \to I^3(k)$ is in fact an isomorphism, but I could not prove this. This question is closely connected with some unproved conjectures of Milnor on the Witt ring (see [9]).) Finally we obtain a canonical epimorphism $\varphi : Um_3(A)/SL_3(A) \to I^3(k)$ and one can check that for any $\alpha, \beta, \gamma \in k^*$ we have

$$\varphi((1-\alpha)X + \alpha,\ (1-\beta)Y + \beta,\ (1-\gamma)Z + \gamma) = \langle 1, -\alpha\rangle\langle 1, -\beta\rangle\langle 1, -\gamma\rangle .$$

The right-hand side of this formula is a three-fold Pfister form which we shall denote by $\langle\langle \alpha, \beta, \gamma\rangle\rangle$. Using the main theorem of Elman-Lam [5] on Pfister forms we obtain

THEOREM 7. ([20]) _The following three conditions are equivalent:_

1) _The unimodular rows_ $((1-\alpha_0)X + \alpha_0,\ (1-\beta_0)Y + \beta_0,\ (1-\beta_0)Z + \beta_0)$ _and_ $((1-\alpha_1)X + \alpha_1,\ (1-\beta_1)Y + \beta_1,\ (1-\gamma_1)Z + \gamma_1)$ _lie in the same orbit relative to the action of the group_ $SL_3(A)$;

2) _The quadratic forms_ $\langle\langle \alpha_0, \beta_0, \gamma_0\rangle\rangle$ _and_ $\langle\langle \alpha_1, \beta_1, \gamma_1\rangle\rangle$ _are isometric;_

3) _The elements_ $\ell(\alpha_0)\cdot\ell(\beta_0)\cdot\ell(\gamma_0)$ _and_ $\ell(\alpha_1)\cdot\ell(\beta_1)\cdot\ell(\gamma_1)$ _of the Milnor group_ $K_3(k)$ _are congruent modulo_ $2\cdot K_3(k)$.

In particular, the following statements are equivalent:

) A^2 satisfies the cancellation condition;

) $I^3(k) = 0$;

) $K_3(k)$ is a 2-divisible group.

Let us look at one particular case. Suppose that the field k s just the field $\mathbb{R}$ of real numbers. Then it is well-known that $(\mathbb{R}) = \mathbf{Z}$, $I(\mathbb{R}) = 2\cdot\mathbf{Z}$ and hence $I^3(\mathbb{R}) = 8\mathbf{Z} \cong \mathbf{Z}$. Moreover, it may be hecked in this case that the epimorphism $G(\mathbb{R}) \to I^3(\mathbb{R}) \cong \mathbf{Z}$ contructed above is really an isomorphism and hence $Um_3(A_{\mathbb{R}})/SL_3(A_{\mathbb{R}}) \cong \mathbf{Z}$. he invariant $\varphi: Um_3(A_{\mathbb{R}}) \to Z$ has a clear topological meaning: enote by Γ the surface in $\mathbb{R}^3$ given by the equation $X^2-X)(Y^2-Y)(Z^2-Z) = 0$. Then Γ has the homotopy type of a twophere and any unimodular row $v \in Um_3(A_{\mathbb{R}})$ defines a continuous mapping $\Gamma \to \mathbb{R}^3-0$ and hence determines an element of the homotopy group $_2(\mathbb{R}^3-0) = \mathbf{Z}$. It may be checked that this topological invariant concides with the algebraic one constructed above.

It is clear that $I^3(k) \neq 0$ for formally real fields. Moreover, sing homomorphisms associated with a discrete valuation on a field, t may be proved that $I^3(k) \neq 0$ in any of the following cases:

) k has a subfield over which k is finitely generated and has ranscendence degree not less than three;

) k is finitely generated, has finite characteristic different rom two, and has transcendence degree over the prime field at least wo;

) k is finitely generated and transcendental over $\mathbb{Q}$.

Even in the cases when $I^3(k) = 0$ one can use the previous esults to construct modules which do not satisfy the cancellation ondition.

HEOREM 8. a) If $I^2(k) \neq 0$, then there exist three-dimensional ffine k-algebras A for which A^2 does not satisfy the cancellation ondition.

b) <u>If</u> k <u>is not quadratically closed, then there exist four-dimensional affine</u> k-<u>algebras</u> A <u>for which</u> A^2 <u>does not satisfy the cancellation condition.</u>

c) <u>For any</u> k <u>there exist five-dimensional affine</u> k-<u>algebras</u> A <u>for which</u> A^2 <u>does not satisfy the cancellation condition.</u>

Using the last part of this statement we can construct for any k a quite concrete five-dimensional k-algebra A and a quite concrete projective A-module P which is stably isomorphic but not isomorphic to A^2. Namely, take $A = k[X_1,\dots,X_6]/(X_1X_4+X_2X_5+X_3X_6-1)$. Then $v = (x_1,x_2,x_3) \in Um_3(A)$ and hence defines a projective A-module P such that $P \oplus A \cong A^3$. If P were isomorphic to A^2, then it would follow that for any k-algebra B the module B^2 satisfies the cancellation condition. Thus by the previous theorem, $P \not\cong A^2$. For char $k \neq 2$, a direct proof of this fact was given earlier by Raynaud [15] using methods of algebraic geometry.

III. The stable rank of polynomial rings.

The last theme that I intend to discuss here is the stable rank of polynomial rings and affine algebras. The stable rank of a ring is one of its most important invariants related to the cancellation problem and to other problems of stabilization in algebraic K-theory. This concept first appeared in the paper of Bass [1] on stabilization for K_1 and the term was introduced by Vaserstein [26]. The <u>stable rank</u> of the ring A (denoted by s.r.(A)) is the least number n for which the following condition is satisfied: Given a unimodular row $(a_0,\dots,a_n) \in Um_{n+1}(A)$, there exist elements $b_1,\dots,b_n \in A$ such that the row of length n $(a_1+a_0b_1,\dots,a_n+a_0b_n)$ is also unimodular. Informally speaking, the stable rank is the least n such that n elements in general position generate the unit ideal. Stable rank is a sort of dimension for a ring. For example, if X is a topological space, Vaserstein [26] has proved that $s.r.(A) = 1+\dim X$ for A the ring of continuous functions on X. The connection between stable rank and more traditional concepts of dimension is given by the following inequality due to Bass: If A is a commutative noetherian ring, then $s.r.(A) \leq 1 + \dim \operatorname{Max} A$. For example, $s.r.(k[X_1,\ldots,X_n]) \leq n+1$. From the geometric point of view, the strict inequality $s.r.(k[X_1,\ldots,X_n]) < n+1$ would signify that if V is any hypersurface in the affine space $\mathbb{A}_k{}^n$ and $f : V \to \mathbb{A}_k{}^n - 0$ is any morphism, then f can be extended to a morphism $\mathbb{A}_k{}^n \to \mathbb{A}_k{}^n - 0$. In some cases the exact value of the stable rank may be computed with the help of topological considerations. For example, in case $k = \mathbb{R}$ one can take $V = S^{n-1}$, an $(n-1)$-sphere, and for f the standard imbedding of S^{n-1} in $\mathbb{R}^n - 0$, then f is not homotopic to a constant and hence cannot be extended to the whole of $\mathbb{R}^n$. Thus $s.r.(\mathbb{R}[X_1,\ldots,X_n]) = n+1$. In the previous argument we used known information about $\pi_{n-1}(\mathbb{R}^n - 0) = \pi_{n-1}(S^{n-1})$. I hope that it will be seen from what I'll say below that for an arbitrary field k Milnor's group $K_n(k)$ plays a role somewhat analogous to that of $\pi_{n-1}(S^{n-1})$ in the case of the field of real numbers. I should also mention that for the field $\mathbb{C}$ topological methods give only a rather weak inequality $s.r.(\mathbb{C}[X_1,\ldots,X_n]) \geq 1 + \frac{n}{2}$.

The exact value of the stable rank of the ring of polynomials in two variables is known in nearly all cases by results of Krusemeyer [6] and Vaserstein (see [27]). Namely, Krusemeyer proved that if $K_2(k) \neq 0$, then $s.r.(k[X_1,X_2]) = 3$. Vaserstein has proved that if k is algebraic over a finite field, then $s.r.(k[X_1,X_2]) = 2$. Moreover by the results of Bass and Tate $K_2(k)$ is equal to zero only for algebraic extensions of finite fields and for some infinite extensions of global fields. The exact value of the stable rank for the latter fields is still unknown. The computation of the stable rank for polynomial rings in three or more variables is based on the following lemma proved in my joint paper with Vaserstein [27]:

Suppose that $r \geq 3$. *Then the following conditions are equivalent*

a) $s.r.(k[X_1,\dots,X_n]) \leq r$;

b) *For any* $(n-1)$-*dimensional affine* k-*algebra* A *and for any ideal* $q \subset A$ *the group of elementary matrices* $E_r(A,q)$ *acts transitively on the set* $Um_r(A,q)$ *of unimodular rows which are congruent to the standard row* $(1,0,\dots,0)$ *modulo* q.

Let A be a commutative ring, q an ideal in A, and $r \geq 2$ a natural number. A mapping $\varphi : Um_r(A,q) \to H$ (an abelian group) is called a *Mennicke symbol* if

1) φ is multiplicative in each variable;

2) $\varphi(a_1,\dots,a_r) = \varphi(a_1,\dots,a_i + ba_j,\dots,a_r)$ $(i \neq j,\ b \in q \text{ if } j = 1)$.

Certainly such a universal Mennicke symbol exists and we shall denote its group of values by $MS_r(A,q)$. For example, if A is a Dedekind domain and $r = 2$, then by the Kubota-Bass theorem (see [3]) $MS_2(A,q) \cong SK_1(A,q)$, but I don't know any invariant description of $MS_r(A,q)$ if $r \geq 3$.

It is easy to see that if q is a principal ideal, then the action of $E_r(A,q)$ on $Um_r(A,q)$ does not change values of Mennicke symbols and so the mapping $Um_r(A,q) \to MS_r(A,q)$ factors through the orbit space $Um_r(A,q)/E_r(A,q)$.

THEOREM 9. ([21]) *Suppose* k *is a field,* $A = k[X_1,\dots,X_n]$ $(n \geq 1)$, *and* q *is the principal ideal in* A *generated by the polynomial* $g = (X_1^2 - X_1)\cdots(X_n^2 - X_n)$. *Then there exists a canonical Mennicke symbol*

$$\varphi : Um_{n+1}(A,q) \to \widetilde{K}_{n+1}(k) = K_{n+1}(k)/\text{Torsion}$$

and the induced homomorphism

$$MS_{n+1}(A,q) \to \widetilde{K}_{n+1}(k)$$

is surjective. If k *is algebraically closed, then* $\widetilde{K}_{n+1}(k) = K_{n+1}(k)$

Bass and Tate) and $MS_{n+1}(A,q) \tilde{\rightarrow} K_{n+1}(k)$. The symbol φ is
haracterized by the following formula:

$$\varphi(p_0,p_1g,\dots,p_ng) =$$

$$\sum_{\substack{{}_1,\dots,p_n \in \mu \in \operatorname{Max} A \\ g \notin \mu}} e_\mu(p_1,\dots,p_n) N_{k(\mu)/k}\left(\ell(p_0(\mu))\cdot \ell\left(\frac{x_1(\mu)}{x_1(\mu)-1}\right)\cdots \ell\left(\frac{x_n(\mu)}{x_n(\mu)-1}\right)\right)$$

if $p_1,\dots,p_n$ have only a finite number of common zeros).

Here e denotes the intersection multiplicity.

It seems rather probable that $MS_{n+1}(A,q) = K_{n+1}(k)$ for rbitrary k (not only for algebraically closed k). This is true or $n=1$ by Krusemeyer's theorem, but to prove this one needs the ransfer homomorphism for Milnor's K-groups. The construction of such transfer was given by Bass and Tate [4], but unfortunately they roved that their definition is unambiguous only modulo torsion; that s the reason why I had to replace $K_{n+1}(k)$ by $\tilde{K}_{n+1}(k)$ in Theorem 9.

Theorem 9 shows that, at least for algebraically closed fields k, ilnor's group $K_{n+1}(k)$ is something like the n-th homotopy group of he space $Um_{n+1}(k) = \mathbb{A}_k^{n+1} - 0$ of unimodular rows of length $(n+1)$ ver k.

OROLLARY. If $n \geq 3$ and $\tilde{K}_n(k) \neq 0$, then $s.r.(k[X_1,\dots,X_n]) = n+1$.

EMARK. The previous corollary is valid for $n = 1,2$ in the following tronger form: For any k, $s.r.(k[x]) = 2$ (obvious). If $K_2(k) \neq 0$, hen $s.r.(k[X_1,X_2]) = 3$ (Krusemeyer).

The Kronecker dimension of a field is its transcendence degree ver the prime subfield, enlarged by one in case of zero character- stic.

THEOREM 10. *Suppose that the Kronecker dimension of a field* k *is equal to* d. *Then* $\mathrm{s.r.}(k[X_1,\ldots,X_n]) = n+1$ *if* $n \leq d$ *and* $[\frac{n+d}{2}]+1 \leq \mathrm{s.r.}(k[X_1,\ldots,X_n]) \leq n+1$ *if* $n \geq d$.

The analogous results are also valid for affine algebras:

THEOREM 11. *If* $\tilde{K}_{n+1}(k) \neq 0$, *then any* n-*dimensional affine* k-*algebra* A *has stable rank equal to* n+1. *In particular, if the Kronecker dimension of* k *is infinite, then* $\mathrm{s.r.}(A) = 1 + \dim A$ *for any affine* k-*algebra* A.

On the other hand we have

THEOREM 12. (Vaserstein). *If* k *is algebraic over a finite field and* A *is an affine* k-*algebra, then* $\mathrm{s.r.}(A) \leq \max(2, \dim A)$.

References

1. H. Bass, K-theory and stable algebra, Inst. Hautes Études Sci. Publ. Math. No. 22 (1964), 5-60.

2. H Bass, Some problems in "classical" algebraic K-theory, pp. 3-73 of Lecture Notes in Math. 342, Springer-Verlag, Berlin and New York, 1973.

3. H. Bass, J. Milnor and J.-P. Serre, Solution of the congruence subgroup problem for SL_n $(n \geq 3)$ and $Sp_{2n}(n \geq 2)$, Inst. Hautes Études Sci. Publ. Math. No. 33 (1967), 59-137 Corrections, *ibid.*, No. 44 (1974), 241-244.

4. H. Bass and J. Tate, The Milnor ring of a global field, pp. 349-446 of Lecture Notes in Math. 342, Springer-Verlag, Berlin and New York, 1973

5. R. Elman and T.-Y. Lam, Pfister forms and K-theory of fields, J. Algebra 23 (1972), 181-213.

6. M.I. Krusemeyer, Fundamental groups, algebraic K-theory, and a

problem of Abhyankar, Invent. Math. 19 (1973), 15-47.

. T.-Y. Lam, Serre's Conjecture, Lecture Notes in Math. 635, Springer-Verlag, Berlin and New York, 1978.

. H. Lindel and W. Lütkebohmert, Projektive Moduln über polynomialen Erweiterungen von Potenzreihenalgebren, Arch. der Math. 28 (1977), 51-54.

. J. Milnor, Algebraic K-theory and quadratic forms, Invent. Math. 9 (1970), 318-344.

0. N. Mohan Kumar, On a question of Bass and Quillen, preprint, Tata Institute, 1976/77

1. M.P. Murthy, Projective A[x]-modules, J. London Math. Soc. 41 (1966), 453-456.

2. M.P. Murthy and R.G. Swan, Vector bundles over affine surfaces, Invent. Math. 36 (1976), 125-165.

3. M. Ojanguren and R. Sridharan, Cancellation of Azumaya algebras, J. Algebra 18 (1971), 501-505.

4. D. Quillen, Projective modules over polynomial rings, Invent. Math. 36 (1976), 167-171.

5. M. Raynaud, Modules projectifs universels, Invent. Math. 6 (1968), 1-26.

6. J.-P. Serre, Modules projectifs et espaces fibrés à fibre vectorielle, Séminaire Dubriel-Pisot, Exposé 23, 1957/58.

7. A.A. Suslin, Projective modules over a polynomial ring are free, Dokl. Akad. Nauk SSSR 229 (1976), 1063-1066 = Soviet Math. Dokl. 17 (1976), 1160-1164.

8. A.A. Suslin, A cancellation theorem for projective modules over algebras, Dokl. Akad. Nauk SSSR 336 (1977), 808-811 = Soviet Math. Dokl. 18 (1977), 1281-1284.

9. A.A. Suslin, On stably free modules, Mat. Sb. 102 (144) (1977), 537-550.

0. A.A. Suslin, Orbits of the group SL_3 and quadratic forms,

preprint LOMI, Steklov Institute, 1977

21. A.A. Suslin, Reciprocity laws and stable range in polynomial rings, preprint LOMI, Steklov Institute, 1978.

22. R.G. Swan, Vector bundles and projective modules, Trans. Amer. Math. Soc. 105 (1962), 264-277.

23. R.G. Swan, Topological examples of projective modules, Trans. Amer. Math. Soc. 230 (1977), 201-234.

24. R.G. Swan, Projective modules over Laurent polynomial rings, Trans. Amer. Math. Soc. 237 (1978), 111-120.

25. R.G. Swan and J. Towber, A class of projective modules which are nearly free, J. Algebra 36 (1975), 427-434.

26. L.N. Vaserstein, Stable rank of rings and dimensionality of topological spaces, Funkcional. Anal. i Priložen. 5 (1971), 17-27 = Functional Anal. Appl. 5 (1971), 102-110.

27. L.N. Vaserstein and A.A. Suslin, The problem of Serre on projective modules over polynomial rings and algebraic K-theory, Izv. Akad. Nauk SSSR Ser. Mat. 40 (1976), 993-1054 = Math. USSR-Izv. 10 (1976), 937-1001.

LOMI
Fontanka 27
Leningrad, USSR

MODULES OVER FULLY BOUNDED NOETHERIAN RINGS

Robert B. Warfield, Jr.

The well-known Forster-Swan theorem gives a bound on the number of generators a finitely generated module over a suitable ring in terms of local data. In sec- n 2 of this paper, we prove a generalization of this result which applies, in ticular, to modules over right Noetherian right fully bounded rings of finite Krull ension. In the first section we give a review of some of the history of the problem, its connection with recent generalizations in a noncommutative setting of Serre's orem and the Bass cancellation theorem. In section 3 we make some brief comments possible extensions to rings which do not satisfy the boundedness conditions. In fourth section we introduce the notion of the stable number of generators of a ule, which makes a connection between the concerns of this paper and the notion of stable range of a ring. The work reported here was supported in part by a grant m the National Science Foundation.

From vector bundles to the Forster-Swan theorem. The questions we are interested here originate in the theory of vector bundles. (All details on vector bundles can found in [8].) If X is a compact Hausdorff space, then a real vector bundle ξ r X is (roughly) a larger space E with a projection $\pi : E \to X$ of E onto X, h that for each $x \in X$, $\pi^{-1}(x)$ is a finite dimensional real vector space, and such t for each $x \in X$, there is a neighborhood U of x such that $\pi^{-1}(U)$ looks like V, for some finite dimensional vector space V. The bundle ξ is trivial if it isomorphic to $X \times V$. There is a natural direct sum in the category of vector bun- s, called the Whitney sum, which we will not define but will illustrate by example:

If M is a compact C^∞ manifold of dimension d immersed in Euclidean space of dimension $d+k$, then we have two vector bundles naturally defined -- the tangent bundle T of M and the normal bundle N (associating to each point the fibre consisting of normal vectors to M at that point). The sum of these two, written $T \oplus N$ is just the trivial bundle, $M \times (R^{d+k})$, which we can think of as the restriction to M of the tangent bundle of the Euclidean space. A basic fact is that if X is a compact Hausdorff space, and ξ is any vector bundle on X, then there is a bundle η such that $\xi \oplus \eta$ is trivial, (see e.g. [14]).

The connection with modules over rings arises in this way. If $C(X)$ is the ring of continuous real-valued functions on X, then for any vector bundle ξ, the continuous sections of ξ, $\Gamma(\xi)$, forms a $C(X)$-module. In particular, if ξ is trivial and of dimension n, then $\Gamma(\xi) \cong (C(X))^n$. Hence, for every vector bundle we get a finitely generated projective module, and this in fact, gives an equivalence of categories between finitely generated projective $C(X)$-modules and vector bundles over the compact Hausdorff space X ([14]).

We now look at several things about a vector bundle, and see how they can be expressed in the language of modules, so that statements about them might make sense for modules over other rings. First, if ξ is a vector bundle of dimension n, (i.e., all the fibers are n-dimensional) what about the corresponding projective module remembers this number n? The point here is that the maximal ideals of $C(X)$ are all determined by points of X, where the maximal ideal M_x corresponding to x is $\{f \in C(X): f(x) = 0\}$. Clearly, $C(X)/M_x \cong R$ (where R is the ring of real numbers), where the isomorphism is given by $f \to f(x)$. Similarly, if $s \in \Gamma(\xi)$, where ξ is the bundle given by $\pi\colon E \to X$, then the map $s \to s(x)$ gives an isomorphism $\Gamma(\xi)/\Gamma(\xi)M_x \to \pi^{-1}(x)$, where $\pi^{-1}(x)$ is an n-dimensional real vector space. In different language, the dimension of the vector bundle is the local number of generators -- the number of generators of $\Gamma(\xi)$ modulo a maximal ideal. Most bundles we think of are over connected spaces, in which case this number is independent of the maximal ideal chosen.

Secondly, if ξ is a vector bundle, and ξ is a summand of a trivial bundle of dimension n, what does this say about the module $\Gamma(\xi)$? This is straightforward --

says precisely that as a module, $\Gamma(\xi)$ is generated by at most n elements.

Finally, if ξ is a bundle, then ξ has a section s which is everywhere non-
-o if and only if $\xi = \eta \oplus \tau$, where τ is a trivial one-dimensional bundle. At the
dule level, this means that $\Gamma(\xi)$ has a summand isomorphic to $C(X)$.

Now if we assume for simplicity that X is a finite simplicial complex of dimen-
on d, then a classical theorem asserts that a bundle of dimension greater than d
ways has an everywhere nonzero section. Anotner classical result asserts that if
and ξ' are bundles, and τ is a trivial bundle, and $\xi \oplus \tau \cong \xi' \oplus \tau$ and the dimen-
on of ξ is greater than d, then $\xi \cong \xi'$, [8, pp. 99-100].

This means, first of all, that a "large enough" vector bundle over such an X
a trivial summand. If ξ is a non-zero bundle over the complex X of dimension
and $\xi \oplus \eta$ is trivial, then applying this to the bundle η, we see that if η has
mension at least $d+1$ then $\eta = \eta' \oplus \tau$, for some trivial one-dimensional τ, since
n$(\xi + \eta') \geq d+1$ we can conclude that $\xi \oplus \eta'$ is again trivial, of dimension less than
fore. We conclude that if X is a complex of dimension d, and ξ a vector bundle
dimension n, then $\Gamma(\xi)$ as a module is generated by at most $n+d$ elements.

Now if we want to find some natural class of rings where analogous theorems might
ld, the ring $C(X)$ seems to be a poor guide, since most rings one studies don't
ok at all like it. However, if X is, say, an irreducible real affine algebraic
riety of dimension d, then the rational functions on X, (the coordinate ring of
, is a Noetherian ring of Krull dimension d, and this suggests that there might
some general theorems about Noetherian rings, where the Krull dimension corresponds
the dimension of X. In the case of commutative Noetherian rings, these are, of
urse, Serre's theorem and the Bass cancellation theorem, both of which have been
tensively generalized in a noncommutative setting by Stafford in [12] and [13].

We are concerned in this paper with the consequence of these theorems, which for
commutative ring of Krull dimension d says that a finitely generated projective
dule locally generated by n elements is actually generated by at most $n+d$ ele-
nts. Now this would be a useful sort of thing for modules which are not projective,
or ideals, for example), and a generalization was obtained by Forster [3], who showed
at if R is a commutative Noetherian ring of Krull dimension d, and A is a

finitely generated module such that for every maximal ideal M of R, A/AM is generated by at most n elements (as a vector space over R/M), then A can be generated by $n+d$ elements. There are a number of generalizations of this around. Let us state one here which applies to non-commutative rings which are not necessarily finite algebras over commutative rings.

If A is a module, we let $g(A)$ be the smallest number which is the cardinality of a set of generators for A.

Theorem A. If R is a Noetherian, right fully bounded ring, of finite Krull dimension d, and A is a finitely generated right R-module, Then

$$g(A) \leq d + \max \{g(A/AM)\}$$

where the maximum is taken over all maximal ideals M of R.

Here the Krull dimension can be taken as either the usual classical Krull dimension defined in terms of prime ideals, or the non-commutative Krull dimension, since these agree for these rings [9]. If R is commutative, $g(A/AM) = g(A_M)$, where A_M is viewed as a module over the localization R_M, so that this result then reduces to Forster's theorem. We remind the reader that in a noncommutative ring, an ideal is prime if for ideals I and J, $IJ \leq P$ implies $I \leq P$ or $J \leq P$, and that a ring is prime if (0) is a prime ideal. A prime ring R is right bounded if for every essential right ideal I of R, there is a (two-sided) ideal J, $J \neq 0$, such that $I \geq J$. A ring R is right fully bounded if R/P is bounded for every prime ideal P. If a property of this sort is referred to without the prefix "right" or "left", then it is intended to apply on both sides. The class of fully bounded Noetherian rings includes finite algebras over commutative Noetherian rings, and also Noetherian P.I. rings. An important feature of these rings is that for any maximal ideal M, R/M is simple Artinian, and for every simple module S, $\mathrm{Ann}(S)$ is a maximal ideal. Therefore, if R is a fully bounded Noetherian ring, A a finitely generated right R-module, and M a maximal ideal of R, then A/AM is a semi-simple module, so the above result is very similar to what one would expect from the commutative case.

learly, maximal ideals are not what one wants to look at if one wants a generaliza-on to Noetherian rings which are not fully bounded.)

In [15], Swan improved Forster's theorem to include finite algebras over mmutative Noetherian rings, and he replaced the Krull dimension by the J-dimension, d the estimate above by a more refined estimate depending on the behavior of the dule A at non-maximal primes. To formulate an analogue of such a result for lly bounded Noetherian rings, we first recall that an ideal in such a ring which is intersection of maximal right ideals is actually the intersection of maximal ideals, nce right primitive factor rings of a right Noetherian right fully bounded ring are tinian. An ideal which is the intersection of maximal ideals is called a J-ideal, d a prime ideal which is a J-ideal will be called a J-prime. The J-dimension R is then the maximal length of a chain of J-primes, and our result will apply enever this is finite.

In the commutative case, one also considers the number of generators of the dules A_P for various primes P. Without localization, we must again replace this th something else. In the commutative case, this number is not the number of gene-tors of A/AP, but rather is the dimension of the vector space $(A/AP)\otimes K$, where is the quotient field of R/P. In our situation, for any prime ideal P, R/P ain has a classical right quotient ring which is Artinian (by Goldie's theorem), ich we will denote by $Q(R/P)$. We therefore define

$$g(P,A) = g((A/AP) \otimes Q(R/P)),$$

ere the module on the right is regarded as a $Q(R/P)$-module. (Alternatively, $g(P,A)$ the minimum cardinality of a subset X of A such that if $[X]$ is the submodule nerated by X, then $A/(AP+[X])$ is singular as an R/P-module.) we let

$$b(P,A) = g(P,A) + \text{J-dim}(R/P),$$

enever $g(P,A)$ is not zero, and $b(P,A) = 0$ otherwise.

eorem B. If R is a right Noetherian, right fully bounded ring, of finite J-mension, and A is a finitely generated right R-module, then

$$g(A) \le \max\{b(P,A)\},$$

where the maximum is taken over all J-primes P.

In the next section, we will give a proof of Theorem B, in slightly more generality than stated above. This theorem is a special case of a more general result which will appear in [17]. However, the result in [17] is much more complicated to prove, and involves many considerations totally irrelevant to Theorem B. (We remark, however that one improvement in [17] is that the ring itself need not have finite J-dimension the only restriction being that Max {b(P,A)} is finite.)

We will not give a proof of Theorem A in this paper. It might seem that it would follow easily from Theorem B, but this is not quite the case. Note, in particular, that in Theorem A, we require the ring R to be Noetherian, while in Theorem B we only require it to be right Noetherian. In the commutative case, (or for algebras over a commutative ring, in which the localization takes place with respect to primes in the commutative ring), it is clear that if P is a prime ideal and M is a maximal ideal containing P, then $g(P,A) \le g(M,A)$, since $A_P = (A_M)_P$. In the noncommutative case, on the other hand, it is easy to see that there can be a prime P with $g(P,A) \ne 0$ and a maximal ideal M containing P such that $g(M,A) = 0$. It is therefore not obvious that from Theorem B we can conclude that A can be generated by $\max\{g(M,A)\} + \text{J-dim}(R)$ elements. In [17], it is proved that if the hypotheses of Theorem A hold, then for any J-prime P, there is a maximal ideal $M \ge P$ with $g(M,A) \ge g(P,A)$, which is precisely what is needed to prove Theorem A.

2. <u>The proof of Theorem B</u>. As in the commutative case, Theorem B can be proved in somewhat greater generality, (replacing "Noetherian" by "J-Noetherian" for example), at the expense of some additional terminology. We will restrict our attention to rings for which right primitive factors are Artinian. In this situation it is reasonable to introduce the previously mentioned notions of J-ideals, J-primes, and J-dimension. A ring is J-Noetherian if it satisfies the ascending chain condition on J-ideals. We will say that R is <u>right J-fully bounded</u> if for every J-prime P,

d every essential right ideal I of R/P, there is a nonzero two-sided ideal of
P contained in I. We summarize our standing hypotheses in the following condition, ich we will simply call (*):

) R is a J-Noetherian, right J-fully bounded ring, of finite J-dimension, for which right primitive factors are Artinian, such that for every J-prime P, R/P is a right Goldie ring.

this situation, it is easy to verify [16] that if I is a J-ideal, then R/I a semiprime right Goldie ring, and, in particular, that there are only a finite mber of primes minimal over I.

All modules will be right modules. If M is an R-module and $X \subseteq M$, we let] be the submodule generated by X and r(X) the right annihilator of X -- those R such that for all $x \in X$, $xr = 0$. An R-module M is singular if for all $x \in M$, x) is an essential right ideal of R. If R is a right Goldie semiprime ring then ery module M has a unique maximal singular submodule Z(M) such that M/Z(M) has nonzero singular submodules. We say M is nonsingular iff $Z(M) = 0$. We refer to] and [5] for details about Goldie rings and nonsingular modules.

mma 1. If R is a ring satisfying (*) and A a finitely generated right R-module, en there are only a finite number of J-primes P at which b(P,A) takes its maxim value.

oof. We will first prove that if P is a J-prime, there is a finite set, Y, J-primes such that if $Q \in Y$, then $Q \supseteq P$, $Q \neq P$, and such that if S is any prime containing P with $g(S,A) > g(P,A)$, then for some $Q \in Y$, $S \supseteq Q$. To prove is, we let P be a J-prime, A a finitely generated module, and $g(P,A) = k$. There , then, a homomorphism $f : R^k \to A$ such that if $f^* : (R/P)^k \to A/AP$ is the induced momorphism, then $C = \text{Coker}(f^*)$ is singular as an R/P-module. The hypothesis (*) early implies that the right annihilator of C is an ideal properly containing P. call this ideal I. The set Y we want consists of those J-primes of R which are minimal over I. The hypotheses clearly imply that this is a finite set. To see that this set works, it suffices to show that if Q is a J-prime, $Q \supseteq P$, and Q does not contain I, then $g(Q,A) \leq g(P,A)$. The homomorphism f induces a short exact sequence

$$(R/Q)^k \to A/AQ \to C/CQ \to 0.$$

If Q does not contain I, then $Q + I > Q$, so as an R/Q-module, C/CQ has a non-zero annihilator. This implies that C/CQ is singular as an R/Q-module, which shows that $g(Q,A) \leq k$, as desired.

We now show by induction on J-dim R that the fact we have just proved implies Lemma 1. We obtain from the previous argument a finite set Y of nonminimal J-primes such that if k is the maximal value of $b(P,A)$ for all minimal J-primes P, then for every J-prime S, $b(S,A) \geq k$ only if for some $Q \in Y$, $S \supseteq Q$. Let $I = \cap_{Q \in Y} Q$. Since J-dim $R/I <$ J-dim R, we see that there are only a finite number of J-primes S, $S \supseteq I$, at which $b(S,A/AI)$ takes its maximal value. Since for these primes, $b(S,A/AI) = b(S,A)$, and the J-primes S at which $b(S,A)$ is maximal are either minimal J-primes or contain I, the result is proved.

Lemma 2. Let A be a module over a right Goldie prime ring R, $x \in A$, and B a submodule of A such that $A/(xR+B)$ is singular. Then there is an element $y \in B$ such that either $A/(x+y)R$ is singular or $(x+y)R \simeq R$.

Proof. It will clearly suffice to prove this under the additional hypothesis that A be nonsingular. In this case, we can find a submodule B' of B such that $B' \cap xR = 0$ and $xR + B'$ is essential in A. [If B' is chosen to be maximal in B with respect to the property that $B' \cap xR = 0$, then clearly $B/[B\cap(B' + xR)]$ is singular, and since A is nonsingular, it follows that $xR + B'$ is essential in $xR + B$.] We may therefore assume that $B \cap xR = 0$. In this case, by induction on the uniform rank of xR, it will suffice to show that either (i) $xR \simeq R$, or (ii) $B = 0$, or (iii) there is an element $y \in B$ such that $r(x+y) < r(x)$. Suppose, then, that (iii) is false, so that no $y \in B$ exists with $r(x+y) < r(x)$. Since $B \cap xR = 0$,

$+y) = r(x) \cap r(y)$ for any $y \in B$, so this means that $r(B) \supseteq r(x)$. Since $r(B)$ is ideal in the prime ring R, it is either essential or zero. If $r(B) = 0$, then $) = 0$, and we are in situation (i). Otherwise, $r(B)$ is essential, so B is a gular submodule of A, and hence zero (since we are assuming A to be nonsingular), we are in situation (ii). In either event, the lemma is proved.

finition. If R is a ring, P is a prime ideal such that R/P is right Goldie, a finitely generated right R-module, and $x \in A$, then x is basic at P if either $,A/xR) = 0$ or $g(P,A/xR) < g(P,A)$.

ma 3. Let R be a ring, X a finite set of prime ideals of R such that if X then R/P is right Goldie, and A a finitely generated right R-module. Then re is an element $x \in A$ which is basic at P for every $P \in X$.

oof. We let $X = \{P_1, \ldots, P_n\}$, where P_n is minimal in the family X, and we sume by induction on n that there is an element $x \in A$ which is basic at each of primes P_i, $i = 1, \ldots, n-1$. The hypothesis on P_n assures us that $P_1 \cap \ldots \cap P_{n-1}$ not contained in P_n, so if $I = P_n + (P_1 \ldots P_{n-1})$, then $I > P_n$. It follows that P_n is essential in R/P_n. If $A' = A/AP_n$, then $A'/A'I$ is singular as an R/P_n-dule, so if we let $B = A'I$, the hypotheses of lemma 2 are satisfied, from which it llows that there is a $y \in A(P_1 \cap \ldots \cap P_{n-1})$ such that $x+y$ is basic at P_n. Since $1 \le i < n$, $y \in AP_i$, it follows that $x+y$ is basic at P_i, since x is, so $x+y$ tisfies the conditions of the lemma.

eorem B'. If R is a ring satisfying hypothesis (*) and A is a finitely generated ght R-module, then

$$g(A) \le \max \{b(P,A)\},$$

ere the maximum is taken over all J-primes P.

oof. The proof is now identical with that in [15, p. 321]. Let X be the set of

J-primes at which $b(P,A)$ is maximal, and let this value be k. By lemma 1 this is a finite set, so by lemma 3 there is an $x \in A$ basic at each of these primes. If $B = A/xR$, then we see easily that for all J-primes P, $b(P,B) < k$. By induction on k, we may assume that B can be generated by a set of $k-1$ or fewer elements, and lifting these elements to elements of A, we have, with our original x, a generati set of at most k elements.

3. Noetherian rings which are not fully bounded. Clearly one would like something like the Forster-Swan theorem for Noetherian rings not satisfying the boundedness hypotheses used above. Just what this result ought to be is not quite clear, but speci cases of whatever theorem will emerge are already known. We mention two. The first is implicit in [10].

Theorem C. If R is a right Noetherian simple ring of right Krull dimension n, $n < \infty$, and A is a finitely generated right R-module, then $g(A) \leq g((0),A) + n$.

Theorem D. ([17, Cor. 7.1]). If R is a Goldie prime ring of right Krull dimensi one, and A a finitely generated right R-module, then

$$g(A) \leq \max b'(P,A)$$

where P ranges over all prime ideals, $b'(M,A) = g(M,A)$ for a (nonzero) maximal ide M, and $b'((0),A) = g((0),A) + 1$.

Note that in each case, a slight difference from Theorem B appears in that we cannot ignore a prime P just because $g(P,A) = 0$.

To give an example to illustrate why these theorems might be interesting from other points of view, we consider briefly the question of rings of bounded module typ -- i.e. rings for which there is an upper bound on the number of generators required for an indecomposable finitely generated module. (Equivalently, in the Noetherian case, which is what concerns us here, these are precisely those rings Morita equivale to a ring over which all finitely generated modules are direct sums of cyclics.)

ollary to Theorem C. If R is simple Noetherian, and n is an upper bound to :h the left and right Krull dimensions of R, (where $n < \infty$), then R is of unded module type, and a finitely generated indecomposable right R-module is gene-:ed by at most $2n+1$ elements.

oof. This is equivalent to a fact pointed out by Stafford in [11], but we give an gument, to illustrate how the above considerations can be applied. If A is a itely generated right R-module, then Stafford's version of Serre's theorem for nple rings [11, 4.3] implies that if A/Z(A) has rank bigger than n+1, then Z(A) has a summand isomorphic to R_R. This clearly can't happen if A is inde-nposable, so we conclude that if A is indecomposable, A/Z(A) has rank at most 1. Since the rank is g((0),A), it follows from Theorem C that $g(A) \leq 2n+1$.

We would like to state a similar corollary for hereditary Noetherian prime (HNP) ngs, but the result leads us into an unresolved problem in the theory of these rings, d therefore gets somewhat complicated. We will therefore just discuss the question which HNP rings are of bounded module type. From [6, Theorem 35], it is clear that R is an HNP ring which does not have enough invertible ideals, then R has in-composable right ideals requiring arbitrarily large numbers of generators. It follows at an HNP ring of bounded module type must necessarily have enough invertible ideals. this case, every finitely generated module is a direct sum of cyclic modules and ght ideals, so it suffices to worry about the number of generators of a right ideal. om Theorem D above, we know that it will suffice to find a bound on the number of nerators of modules of the form A/AM, where A is a right ideal and M a maximal eal. According to [6, Theorem 35], if A is an essential right ideal of R and I maximal invertible ideal, then

$$\text{length } (A/AI) = \text{length } (R/I).$$

particular, if a maximal ideal M is invertible, then A/AM is cyclic, for any ght ideal A. Clearly if, for example, there were a bound on the lengths of the fac-rs R/I for maximal invertible ideals I which are not maximal ideals, then it would follow that R was of bounded module type. No examples are known in which there are more than a finite number of such ideals I, so the obvious conjecture is that the HNP rings of bounded module type are precisely the HNP rings with enough invertible ideals.

4. Stable generation of modules. It was discovered by Eisenbud and Evans [2, Theorem B] that the estimate of the number of generators of a module given by the Forster-Swan theorem is actually an estimate of something else. If A is a finitely generated right R-module, we say that A is stably generated by n elements if given any set of generators $\{x_1,\ldots,x_t\}$ $(t>n)$, there are elements $y_i \in (x_{n+1}R+\ldots+x_tR)$, $(1\le i\le n)$ such that the elements x_i+y_i, $(1\le i\le n)$, generate A. We let s(A) be the smallest integer n such that A is stably generated by n elements, with the understanding that if no such integer exists, then $s(A)=\infty$. As an example, if R is the ring of all linear transformations of an infinite dimensional vector space, then R_R is cyclic, but $s(R_R)=\infty$. However, what Eisenbud and Evans prove in [2] is that for a ring R which is a finite algebra over a commutative ring S which is J-Noetherian of finite J-dimension, the bound on g(A), the number of generators of A, given by the Forster-Swan theorem is also a bound on the stable number of generators, s(A). In [17], we prove the analogous result for right Noetherian right fully bounded rings, (and also a somewhat larger class of rings, so that the result includes the result of Eisenbud and Evans). The commutative methods used by Eisenbud and Evans do not carry over, and the arguments become quite complicated, which is why we wished to present the much clearer proof of the unstable version of the theorem in this paper.

The notion of the stable number of generators of a module arose first in the special case in which we regard the ring as a module over itself. In [1], Bass investigated what is, in our notation, $s(R_R)$. Modifying his terminology, it is now customary to say that n is in the stable range for R if $s(R_R)\le n$. This notion is significant for the study of $K_1(R)$. We recall that $E_n(R)$ is the subgroup of $GL_n(R)$ generated by the elementary transformations (or transvections). One key point about the stable range is that if d is in the stable range for R, then E_n acts ansitively on the unimodular rows of R^n provided that $n \geq d+1$. There is a similar terpretation of $s(A)$, for any module A. We state it as a theorem for emphasis.

eorem E. Let R be a ring, A a finitely generated right R-module, with $s(A) < \infty$, d $n \geq s(A) + g(A)$. Then if $f: R^n \to A$ and $g: R^n \to A$ are two epimorphisms. there is element $e \in E_n(R)$ such that $f = ge$.

The proof of this (though not explicitly so stated) is contained in the first secon of [18]. From our point of view, the fact that we can choose e in $E_n(R)$ is t really the point -- the interesting result is that for large enough n, any two imorphisms $R^n \to A$ are equivalent up to an automorphism of R^n.

Even in situations in which an analogue of the Forster-Swan theorem is not availle, it may still be possible to give an estimate for $s(A)$, and, in particular, to ove that $s(A)$ is finite. It is shown in [18, Theorem 4] that if R is a ring th n in the stable range, and A is any finitely generated right R-module, then

$$s(A) \leq g(A) + n - 1.$$

r example, this applies to modules over any commutative ring of finite classical ull dimension, since Heitman has shown in [7] that a commutative ring with classical ull dimension d has $d+2$ in the stable range. More closely related to our interts in this paper is Stafford's result [11, 2.5] that a right Noetherian right ideal variant ring of (noncommutative) Krull dimension d has $d+1$ in the stable range. should be possible to improve the resulting estimate on $s(A)$ for modules over ese rings, if we knew the right analogue of the Forster-Swan theorem for them. In rticular, this has been done for torsion modules over simple rings in [11, 2.2].

REFERENCES

H. Bass, K-theory and stable algebra, Pub. Math. I.H.E.S. 22 (1964), 5-60.

D. Eisenbud and E. G. Evans, Jr., Generating modules efficiently: theorems from algebraic K-theory, J. Alg. 27 (1973), 278-305.

3. O. Forster, Uber die anzahl der Erzeugenden einen Ideals in einem Noetherian Ring, Math. Zeit. 84 (1964), 80-87.

4. A. W. Goldie, The Structure of Noetherian Rings, In: Lectures on Rings and Modules, Lecture Notes in Mathematics no. 246, Springer-Verlag, Berlin, 1972, pp. 213-321.

5. K. R. Goodearl, Ring Theory: Nonsingular Rings and Modules, Marcel Dekker, New York, 1976.

6. ____________ and R. B. Warfield, Jr., Simple modules over hereditary Noetheria prime rings, J. Alg. (to appear).

7. R. Heitman, Generating ideals in Prufer domains, Pac. J. Math. 62 (1976), 117-1

8. D. Husemoller, Fibre Bundles, McGraw-Hill, New York, 1966.

9. G. Krause, On fully left bounded left Noetherian rings, J. Alg. 23 (1972), 88-9

10. J. T. Stafford, Completely faithful modules and ideals of simple Noetherian ring Bull. London Math. Soc. 8 (1976), 168-173.

11. ____________, Stable structure of noncommutative Noetherian rings, J. Alg. 47 (1977), 244-267.

12. ____________, Stable structure of noncommutative Noetherian rings II, J. Alg. 52 (1978), 218-235.

13. ____________, Cancellation for nonprojective modules, (to appear).

14. R. G. Swan, Vector bundles and projective modules, Trans. Amer. Math. Soc. 105 (1962), 264-277.

15. ____________, The number of generators of a module, Math. Zeit. 102 (1967), 318-322.

16. R. B. Warfield, Jr., Cancellation of modules and groups and the stable range of endomorphism rings, (to appear).

17. ____________, The number of generators of a module over a fully bounded ring, (to appear).

18. ____________, Stable generation of modules, (to appear).

580: C. Castaing and M. Valadier, Convex Analysis and Meas-
le Multifunctions. VIII, 278 pages. 1977.

581: Séminaire de Probabilités XI, Université de Strasbourg.
eedings 1975/1976. Edité par C. Dellacherie, P. A. Meyer et
Veil. VI, 574 pages. 1977.

582: J. M. G. Fell, Induced Representations and Banach
ebraic Bundles. IV, 349 pages. 1977.

583: W. Hirsch, C. C. Pugh and M. Shub, Invariant Manifolds.
9 pages. 1977.

584: C. Brezinski, Accélération de la Convergence en Analyse
érique. IV, 313 pages. 1977.

585: T. A. Springer, Invariant Theory. VI, 112 pages. 1977.

586: Séminaire d'Algèbre Paul Dubreil, Paris 1975–1976
me Année). Edited by M. P. Malliavin. VI, 188 pages. 1977.

587: Non-Commutative Harmonic Analysis. Proceedings 1976.
ed by J. Carmona and M. Vergne. IV, 240 pages. 1977.

588: P. Molino, Théorie des G-Structures: Le Problème d'Equi-
ce. VI, 163 pages. 1977.

589: Cohomologie l-adique et Fonctions L. Séminaire de
métrie Algébrique du Bois-Marie 1965–66, SGA 5. Edité par
usie. XII, 484 pages. 1977.

590: H. Matsumoto, Analyse Harmonique dans les Systèmes de
Bornologiques de Type Affine. IV, 219 pages. 1977.

591: G. A. Anderson, Surgery with Coefficients. VIII, 157 pages.
.

592: D. Voigt, Induzierte Darstellungen in der Theorie der end-
n, algebraischen Gruppen. V, 413 Seiten. 1977.

593: K. Barbey and H. König, Abstract Analytic Function Theory
Hardy Algebras. VIII, 260 pages. 1977.

594: Singular Perturbations and Boundary Layer Theory, Lyon
. Edited by C. M. Brauner, B. Gay, and J. Mathieu. VIII, 539
s. 1977.

595: W. Hazod, Stetige Faltungshalbgruppen von Wahrschein-
eitsmaßen und erzeugende Distributionen. XIII, 157 Seiten. 1977.

596: K. Deimling, Ordinary Differential Equations in Banach
ces. VI, 137 pages. 1977.

597: Geometry and Topology, Rio de Janeiro, July 1976. Pro-
ings. Edited by J. Palis and M. do Carmo. VI, 866 pages. 1977.

598: J. Hoffmann-Jørgensen, T. M. Liggett et J. Neveu, Ecole
de Probabilités de Saint-Flour VI – 1976. Edité par P.-L. Henne-
. XII, 447 pages. 1977.

599: Complex Analysis, Kentucky 1976. Proceedings. Edited
D. Buckholtz and T. J. Suffridge. X, 159 pages. 1977.

600: W. Stoll, Value Distribution on Parabolic Spaces. VIII,
pages. 1977.

601: Modular Functions of one Variable V, Bonn 1976. Proceedings.
ed by J.-P. Serre and D. B. Zagier. VI, 294 pages. 1977.

602: J. P. Brezin, Harmonic Analysis on Compact Solvmanifolds.
179 pages. 1977.

603: B. Moishezon, Complex Surfaces and Connected Sums of
plex Projective Planes. IV, 234 pages. 1977.

604: Banach Spaces of Analytic Functions, Kent, Ohio 1976.
eedings. Edited by J. Baker, C. Cleaver and Joseph Diestel. VI,
ages. 1977.

605: Sario et al., Classification Theory of Riemannian Manifolds.
498 pages. 1977.

606: Mathematical Aspects of Finite Element Methods. Pro-
ings 1975. Edited by I. Galligani and E. Magenes. VI, 362 pages.
.

607: M. Métivier, Reelle und Vektorwertige Quasimartingale
die Theorie der Stochastischen Integration. X, 310 Seiten. 1977.

608: Bigard et al., Groupes et Anneaux Réticulés. XIV, 334
s. 1977.

Vol. 609: General Topology and Its Relations to Modern Analysis and Algebra IV. Proceedings 1976. Edited by J. Novák. XVIII, 225 pages. 1977.

Vol. 610: G. Jensen, Higher Order Contact of Submanifolds of Homogeneous Spaces. XII, 154 pages. 1977.

Vol. 611: M. Makkai and G. E. Reyes, First Order Categorical Logic. VIII, 301 pages. 1977.

Vol. 612: E. M. Kleinberg, Infinitary Combinatorics and the Axiom of Determinateness. VIII, 150 pages. 1977.

Vol. 613: E. Behrends et al., L^p-Structure in Real Banach Spaces. X, 108 pages. 1977.

Vol. 614: H. Yanagihara, Theory of Hopf Algebras Attached to Group Schemes. VIII, 308 pages. 1977.

Vol. 615: Turbulence Seminar, Proceedings 1976/77. Edited by P. Bernard and T. Ratiu. VI, 155 pages. 1977.

Vol. 616: Abelian Group Theory, 2nd New Mexico State University Conference, 1976. Proceedings. Edited by D. Arnold, R. Hunter and E. Walker. X, 423 pages. 1977.

Vol. 617: K. J. Devlin, The Axiom of Constructibility: A Guide for the Mathematician. VIII, 96 pages. 1977.

Vol. 618: I. I. Hirschman, Jr. and D. E. Hughes, Extreme Eigen Values of Toeplitz Operators. VI, 145 pages. 1977.

Vol. 619: Set Theory and Hierarchy Theory V, Bierutowice 1976. Edited by A. Lachlan, M. Srebrny, and A. Zarach. VIII, 358 pages. 1977.

Vol. 620: H. Popp, Moduli Theory and Classification Theory of Algebraic Varieties. VIII, 189 pages. 1977.

Vol. 621: Kauffman et al., The Deficiency Index Problem. VI, 112 pages. 1977.

Vol. 622: Combinatorial Mathematics V, Melbourne 1976. Proceedings. Edited by C. Little. VIII, 213 pages. 1977.

Vol. 623: I. Erdelyi and R. Lange, Spectral Decompositions on Banach Spaces. VIII, 122 pages. 1977.

Vol. 624: Y. Guivarc'h et al., Marches Aléatoires sur les Groupes de Lie. VIII, 292 pages. 1977.

Vol. 625: J. P. Alexander et al., Odd Order Group Actions and Witt Classification of Innerproducts. IV, 202 pages. 1977.

Vol. 626: Number Theory Day, New York 1976. Proceedings. Edited by M. B. Nathanson. VI, 241 pages. 1977.

Vol. 627: Modular Functions of One Variable VI, Bonn 1976. Proceedings. Edited by J.-P. Serre and D. B. Zagier. VI, 339 pages. 1977.

Vol. 628: H. J. Baues, Obstruction Theory on the Homotopy Classification of Maps. XII, 387 pages. 1977.

Vol. 629: W.A. Coppel, Dichotomies in Stability Theory. VI, 98 pages. 1978.

Vol. 630: Numerical Analysis, Proceedings, Biennial Conference, Dundee 1977. Edited by G. A. Watson. XII, 199 pages. 1978.

Vol. 631: Numerical Treatment of Differential Equations. Proceedings 1976. Edited by R. Bulirsch, R. D. Grigorieff, and J. Schröder. X, 219 pages. 1978.

Vol. 632: J.-F. Boutot, Schéma de Picard Local. X, 165 pages. 1978.

Vol. 633: N. R. Coleff and M. E. Herrera, Les Courants Résiduels Associés à une Forme Méromorphe. X, 211 pages. 1978.

Vol. 634: H. Kurke et al., Die Approximationseigenschaft lokaler Ringe. IV, 204 Seiten. 1978.

Vol. 635: T. Y. Lam, Serre's Conjecture. XVI, 227 pages. 1978.

Vol. 636: Journées de Statistique des Processus Stochastiques, Grenoble 1977, Proceedings. Edité par Didier Dacunha-Castelle et Bernard Van Cutsem. VII, 202 pages. 1978.

Vol. 637: W. B. Jurkat, Meromorphe Differentialgleichungen. VII, 194 Seiten. 1978.

Vol. 638: P. Shanahan, The Atiyah-Singer Index Theorem, An Introduction. V, 224 pages. 1978.

Vol. 639: N. Adasch et al., Topological Vector Spaces. V, 125 pages. 1978.

Vol. 640: J. L. Dupont, Curvature and Characteristic Classes. X, 175 pages. 1978.

Vol. 641: Séminaire d'Algèbre Paul Dubreil, Proceedings Paris 1976–1977. Edité par M. P. Malliavin. IV, 367 pages. 1978.

Vol. 642: Theory and Applications of Graphs, Proceedings, Michigan 1976. Edited by Y. Alavi and D. R. Lick. XIV, 635 pages. 1978.

Vol. 643: M. Davis, Multiaxial Actions on Manifolds. VI, 141 pages. 1978.

Vol. 644: Vector Space Measures and Applications I, Proceedings 1977. Edited by R. M. Aron and S. Dineen. VIII, 451 pages. 1978.

Vol. 645: Vector Space Measures and Applications II, Proceedings 1977. Edited by R. M. Aron and S. Dineen. VIII, 218 pages. 1978.

Vol. 646: O. Tammi, Extremum Problems for Bounded Univalent Functions. VIII, 313 pages. 1978.

Vol. 647: L. J. Ratliff, Jr., Chain Conjectures in Ring Theory. VIII, 133 pages. 1978.

Vol. 648: Nonlinear Partial Differential Equations and Applications, Proceedings, Indiana 1976–1977. Edited by J. M. Chadam. VI, 206 pages. 1978.

Vol. 649: Séminaire de Probabilités XII, Proceedings, Strasbourg, 1976–1977. Edité par C. Dellacherie, P. A. Meyer et M. Weil. VIII, 805 pages. 1978.

Vol. 650: C*-Algebras and Applications to Physics. Proceedings 1977. Edited by H. Araki and R. V. Kadison. V, 192 pages. 1978.

Vol. 651: P. W. Michor, Functors and Categories of Banach Spaces. VI, 99 pages. 1978.

Vol. 652: Differential Topology, Foliations and Gelfand-Fuks-Cohomology, Proceedings 1976. Edited by P. A. Schweitzer. XIV, 252 pages. 1978.

Vol. 653: Locally Interacting Systems and Their Application in Biology. Proceedings, 1976. Edited by R. L. Dobrushin, V. I. Kryukov and A. L. Toom. XI, 202 pages. 1978.

Vol. 654: J. P. Buhler, Icosahedral Golois Representations. III, 143 pages. 1978.

Vol. 655: R. Baeza, Quadratic Forms Over Semilocal Rings. VI, 199 pages. 1978.

Vol. 656: Probability Theory on Vector Spaces. Proceedings, 1977. Edited by A. Weron. VIII, 274 pages. 1978.

Vol. 657: Geometric Applications of Homotopy Theory I, Proceedings 1977. Edited by M. G. Barratt and M. E. Mahowald. VIII, 459 pages. 1978.

Vol. 658: Geometric Applications of Homotopy Theory II, Proceedings 1977. Edited by M. G. Barratt and M. E. Mahowald. VIII, 487 pages. 1978.

Vol. 659: Bruckner, Differentiation of Real Functions. X, 247 pages. 1978.

Vol. 660: Equations aux Dérivée Partielles. Proceedings, 1977. Edité par Pham The Lai. VI, 216 pages. 1978.

Vol. 661: P. T. Johnstone, R. Paré, R. D. Rosebrugh, D. Schumacher, R. J. Wood, and G. C. Wraith, Indexed Categories and Their Applications. VII, 260 pages. 1978.

Vol. 662: Akin, The Metric Theory of Banach Manifolds. XIX, 306 pages. 1978.

Vol. 663: J. F. Berglund, H. D. Junghenn, P. Milnes, Compact Right Topological Semigroups and Generalizations of Almost Periodicity. X, 243 pages. 1978.

Vol. 664: Algebraic and Geometric Topology, Proceedings, 1977. Edited by K. C. Millett. XI, 240 pages. 1978.

Vol. 665: Journées d'Analyse Non Linéaire. Proceedings, 1977. Edité par P. Bénilan et J. Robert. VIII, 256 pages. 1978.

Vol. 666: B. Beauzamy, Espaces d'Interpolation Réels: Topologie et Géometrie. X, 104 pages. 1978.

Vol. 667: J. Gilewicz, Approximants de Padé. XIV, 511 pages. 1978.

Vol. 668: The Structure of Attractors in Dynamical Systems. Proceedings, 1977. Edited by J. C. Martin, N. G. Markley and W. Perrizo. VI, 264 pages. 1978.

Vol. 669: Higher Set Theory. Proceedings, 1977. Edited by G. H. Müller and D. S. Scott. XII, 476 pages. 1978.

Vol. 670: Fonctions de Plusieurs Variables Complexes III, ceedings, 1977. Edité par F. Norguet. XII, 394 pages. 1978.

Vol. 671: R. T. Smythe and J. C. Wierman, First-Passage Percula on the Square Lattice. VIII, 196 pages. 1978.

Vol. 672: R. L. Taylor, Stochastic Convergence of Weighted Su of Random Elements in Linear Spaces. VII, 216 pages. 1978.

Vol. 673: Algebraic Topology, Proceedings 1977. Edited by P. man, R. Piccinini and D. Sjerve. VI, 278 pages. 1978.

Vol. 674: Z. Fiedorowicz and S. Priddy, Homology of Clas Groups Over Finite Fields and Their Associated Infinite Loop Spa VI, 434 pages. 1978.

Vol. 675: J. Galambos and S. Kotz, Characterizations of Probab Distributions. VIII, 169 pages. 1978.

Vol. 676: Differential Geometrical Methods in Mathematical Physic Proceedings, 1977. Edited by K. Bleuler, H. R. Petry and A. Re VI, 626 pages. 1978.

Vol. 677: Séminaire Bourbaki, vol. 1976/77, Exposés 489–506 264 pages. 1978.

Vol. 678: D. Dacunha-Castelle, H. Heyer et B. Roynette. Ecole d de Probabilités de Saint-Flour. VII-1977. Edité par P. L. Henneq IX, 379 pages. 1978.

Vol. 679: Numerical Treatment of Differential Equations in Appl tions, Proceedings, 1977. Edited by R. Ansorge and W. Törnig. 163 pages. 1978.

Vol. 680: Mathematical Control Theory, Proceedings, 1977. Ed by W. A. Coppel. IX, 257 pages. 1978.

Vol. 681: Séminaire de Théorie du Potentiel Paris, No. 3, Directe M. Brelot, G. Choquet et J. Deny. Rédacteurs: F. Hirsch et G. Mo bodzki. VII, 294 pages. 1978.

Vol. 682: G. D. James, The Representation Theory of the Symme Groups. V, 156 pages. 1978.

Vol. 683: Variétés Analytiques Compactes, Proceedings, 19 Edité par Y. Hervier et A. Hirschowitz. V, 248 pages. 1978.

Vol. 684: E. E. Rosinger, Distributions and Nonlinear Partial ferential Equations. XI, 146 pages. 1978.

Vol. 685: Knot Theory, Proceedings, 1977. Edited by J. C. Hausma VII, 311 pages. 1978.

Vol. 686: Combinatorial Mathematics, Proceedings, 1977. Ed by D. A. Holton and J. Seberry. IX, 353 pages. 1978.

Vol. 687: Algebraic Geometry, Proceedings, 1977. Edited by L Olson. V, 244 pages. 1978.

Vol. 688: J. Dydak and J. Segal, Shape Theory. VI, 150 pages. 19

Vol. 689: Cabal Seminar 76–77, Proceedings, 1976–77. Edited A.S. Kechris and Y. N. Moschovakis. V, 282 pages. 1978.

Vol. 690: W. J. J. Rey, Robust Statistical Methods. VI, 128 pages. 19

Vol. 691: G. Viennot, Algèbres de Lie Libres et Monoïdes Lib III, 124 pages. 1978.

Vol. 692: T. Husain and S. M. Khaleelulla, Barrelledness in To logical and Ordered Vector Spaces. IX, 258 pages. 1978.

Vol. 693: Hilbert Space Operators, Proceedings, 1977. Edited J. M. Bachar Jr. and D. W. Hadwin. VIII, 184 pages. 1978.

Vol. 694: Séminaire Pierre Lelong – Henri Skoda (Analyse) An 1976/77. VII, 334 pages. 1978.

Vol. 695: Measure Theory Applications to Stochastic Analy Proceedings, 1977. Edited by G. Kallianpur and D. Kölzow. 261 pages. 1978.

Vol. 696: P. J. Feinsilver, Special Functions, Probability Semigrou and Hamiltonian Flows. VI, 112 pages. 1978.

Vol. 697: Topics in Algebra, Proceedings, 1978. Edited by M. F. N man. XI, 229 pages. 1978.

Vol. 698: E. Grosswald, Bessel Polynomials. XIV, 182 pages. 19

Vol. 699: R. E. Greene and H.-H. Wu, Function Theory on Manifo Which Possess a Pole. III, 215 pages. 1979.